Succeed

Eureka Math®
Grade 2
Modules 6–8

Published by Great Minds®.

Copyright © 2018 Great Minds®.

Printed in the U.S.A.
This book may be purchased from the publisher at eureka-math.org.
BAB   10   9   8   7   6   5

ISBN 978-1-64054-039-2

G2-M6-M8-S-06.2018

Learn ◆ Practice ◆ Succeed

Eureka Math® student materials for *A Story of Units*® (K–5) are available in the *Learn, Practice, Succeed* trio. This series supports differentiation and remediation while keeping student materials organized and accessible. Educators will find that the *Learn, Practice,* and *Succeed* series also offers coherent—and therefore, more effective—resources for Response to Intervention (RTI), extra practice, and summer learning.

Learn

Eureka Math Learn serves as a student's in-class companion where they show their thinking, share what they know, and watch their knowledge build every day. *Learn* assembles the daily classwork—Application Problems, Exit Tickets, Problem Sets, templates—in an easily stored and navigated volume.

Practice

Each *Eureka Math* lesson begins with a series of energetic, joyous fluency activities, including those found in *Eureka Math Practice.* Students who are fluent in their math facts can master more material more deeply. With *Practice,* students build competence in newly acquired skills and reinforce previous learning in preparation for the next lesson.

Together, *Learn* and *Practice* provide all the print materials students will use for their core math instruction.

Succeed

Eureka Math Succeed enables students to work individually toward mastery. These additional problem sets align lesson by lesson with classroom instruction, making them ideal for use as homework or extra practice. Each problem set is accompanied by a Homework Helper, a set of worked examples that illustrate how to solve similar problems.

Teachers and tutors can use *Succeed* books from prior grade levels as curriculum-consistent tools for filling gaps in foundational knowledge. Students will thrive and progress more quickly as familiar models facilitate connections to their current grade-level content.

Students, families, and educators:

Thank you for being part of the *Eureka Math®* community, where we celebrate the joy, wonder, and thrill of mathematics.

Nothing beats the satisfaction of success—the more competent students become, the greater their motivation and engagement. The *Eureka Math Succeed* book provides the guidance and extra practice students need to shore up foundational knowledge and build mastery with new material.

What is in the Succeed *book?*

Eureka Math Succeed books deliver supported practice sets that parallel the lessons of *A Story of Units®*. Each *Succeed* lesson begins with a set of worked examples, called *Homework Helpers*, that illustrate the modeling and reasoning the curriculum uses to build understanding. Next, students receive scaffolded practice through a series of problems carefully sequenced to begin from a place of confidence and add incremental complexity.

How should Succeed *be used?*

The collection of *Succeed* books can be used as differentiated instruction, practice, homework, or intervention. When coupled with *Affirm®*, *Eureka Math*'s digital assessment system, *Succeed* lessons enable educators to give targeted practice and to assess student progress. *Succeed*'s perfect alignment with the mathematical models and language used across *A Story of Units* ensures that students feel the connections and relevance to their daily instruction, whether they are working on foundational skills or getting extra practice on the current topic.

Where can I learn more about Eureka Math *resources?*

The Great Minds® team is committed to supporting students, families, and educators with an ever-growing library of resources, available at eureka-math.org. The website also offers inspiring stories of success in the *Eureka Math* community. Share your insights and accomplishments with fellow users by becoming a *Eureka Math* Champion.

Best wishes for a year filled with Eureka moments!

Jill Diniz

Jill Diniz
Director of Mathematics
Great Minds

Contents

Module 6: Foundations of Multiplication and Division

Module 7: Problem Solving with Length, Money, and Data

Module 8: Time, Shapes, and Fractions as Equal Parts of Shapes

Grade 2
Module 6

$2 + 2 + 2 = 6$
I can think $2 + 2 = 4$ and $4 + 2 = 6$.

Repeated addition in Grade 2 ...

$3 \times 2 = 6$
I can think 3 groups of 2 equals 6.

leads to multiplication in Grade 3.

By putting the apples into groups of 2, I create 5 equal groups of two apples.

1. Circle groups of two apples.

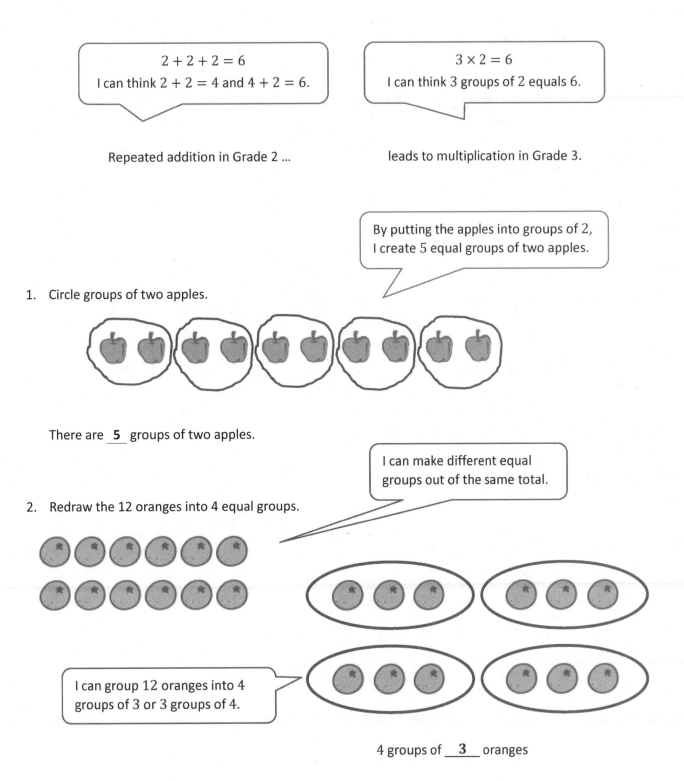

There are __5__ groups of two apples.

I can make different equal groups out of the same total.

2. Redraw the 12 oranges into 4 equal groups.

I can group 12 oranges into 4 groups of 3 or 3 groups of 4.

4 groups of __3__ oranges

EUREKA MATH

3. Redraw the 12 oranges into 3 equal groups.

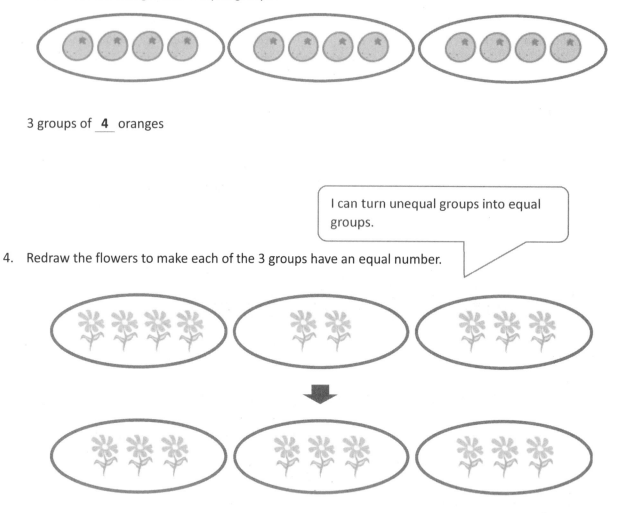

3 groups of __4__ oranges

I can turn unequal groups into equal groups.

4. Redraw the flowers to make each of the 3 groups have an equal number.

3 groups of __3__ flowers = __9__ flowers.

Lesson 1:   Use manipulatives to create equal groups.

Name _____   Date _____

1. Circle groups of two shirts.

There are _____ groups of two shirts.

2. Circle groups of three pants.

There are _____ groups of three pants.

3. Redraw the 12 wheels into 3 equal groups.

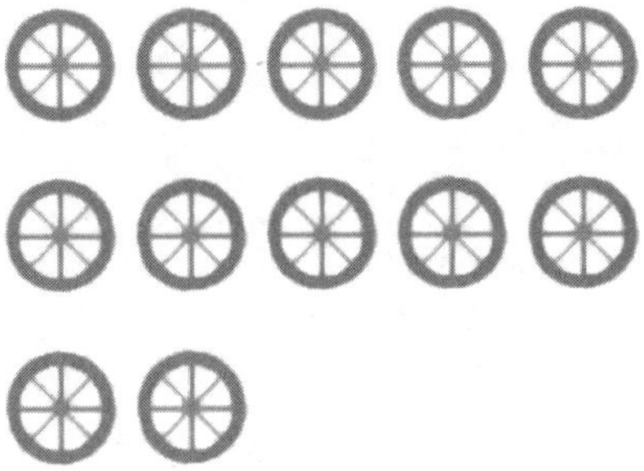

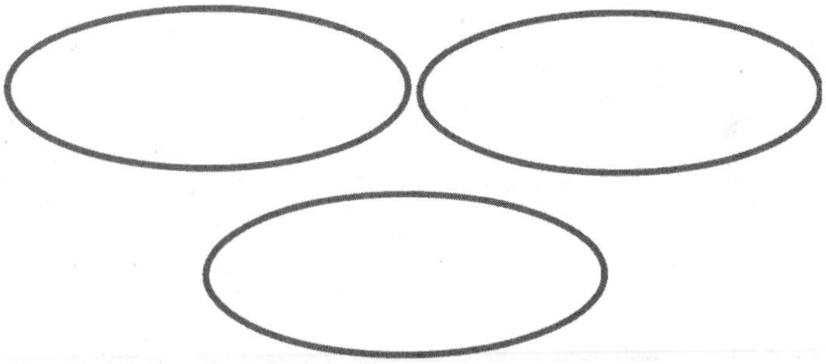

3 groups of _____ wheels

4. Redraw the 12 wheels into 4 equal groups.

4 groups of _____ wheels

5. Redraw the apples to make each of the 4 groups have an equal amount.

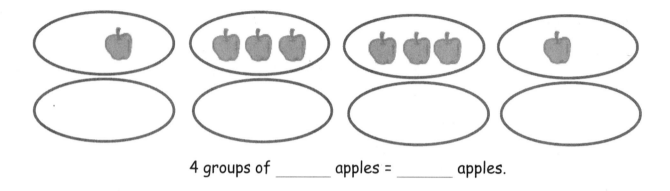

4 groups of _____ apples = _____ apples.

6. Redraw the oranges to make 3 equal groups.

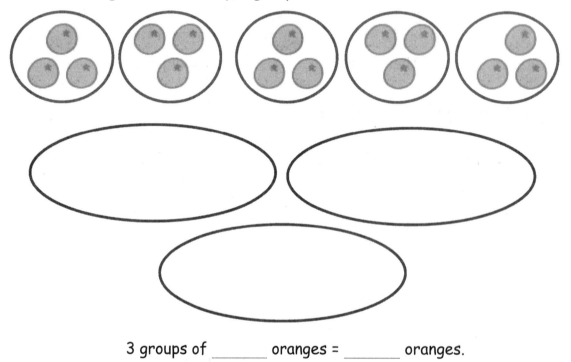

3 groups of _____ oranges = _____ oranges.

1. Write a repeated addition equation to show the number of objects in each group. Then, find the total.

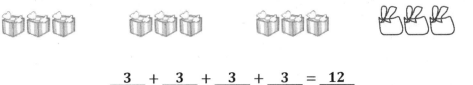

$$\underline{\ 2\ } + \underline{\ 2\ } + \underline{\ 2\ } = \underline{\ 6\ }$$

3 groups of __2__ = __6__

> There are 2 pencils in each group, so the repeated addition sentence is $2 + 2 + 2 = 6$. We can say 3 groups of 2 equals 6.

2. Draw 1 more group of three. Then, write a repeated addition equation to match.

$$\underline{\ 3\ } + \underline{\ 3\ } + \underline{\ 3\ } + \underline{\ 3\ } = \underline{\ 12\ }$$

__4__ groups of 3 = __12__

> When I draw another group of 3 boxes, I have to add another 3 to the repeated addition sentence because now there are 4 groups of 3.

Lesson 2:   Use math drawings to represent equal groups, and relate to repeated addition.

Name _____   Date _____

1. Write a repeated addition equation to show the number of objects in each group. Then, find the total.

   a.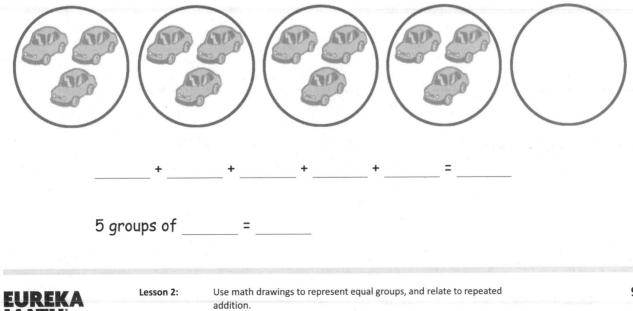

   _____ + _____ + _____ = _____

   3 groups of _____ = _____

   b.

   _____ + _____ + _____ + _____ = _____

   4 groups of _____ = _____

2. Draw 1 more equal group.

   _____ + _____ + _____ + _____ + _____ = _____

   5 groups of _____ = _____

3. Draw 1 more group of four. Then, write a repeated addition equation to match.

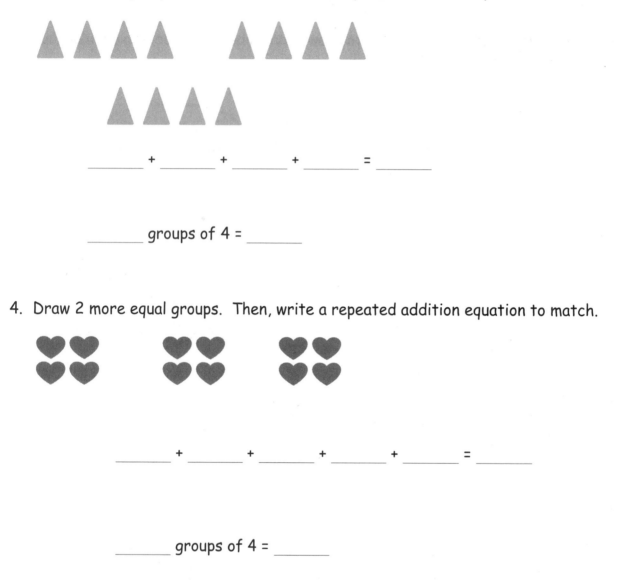

_____ + _____ + _____ + _____ = _____

_____ groups of 4 = _____

4. Draw 2 more equal groups. Then, write a repeated addition equation to match.

_____ + _____ + _____ + _____ + _____ = _____

_____ groups of 4 = _____

5. Draw 4 groups of 3 circles. Then, write a repeated addition equation to match.

Lesson 2:   Use math drawings to represent equal groups, and relate to repeated addition.

© 2018 Great Minds®. eureka-math.org

1. Write a repeated addition equation to match the picture. Then, group the addends into pairs to show a more efficient way to add.

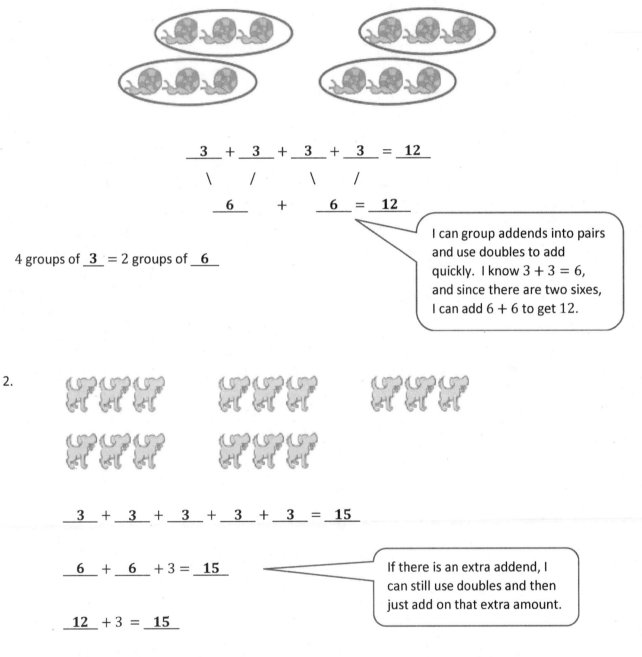

$$\underline{3} + \underline{3} + \underline{3} + \underline{3} = \underline{12}$$
$$\backslash/\backslash/$$
$$\underline{6} \quad + \quad \underline{6} = \underline{12}$$

4 groups of _3_ = 2 groups of _6_

> I can group addends into pairs and use doubles to add quickly. I know $3 + 3 = 6$, and since there are two sixes, I can add $6 + 6$ to get 12.

2.

$$\underline{3} + \underline{3} + \underline{3} + \underline{3} + \underline{3} = \underline{15}$$

$$\underline{6} + \underline{6} + 3 = \underline{15}$$

> If there is an extra addend, I can still use doubles and then just add on that extra amount.

$$\underline{12} + 3 = \underline{15}$$

EUREKA MATH

Lesson 3:   Use math drawings to represent equal groups, and relate to repeated addition.

11

© 2018 Great Minds®. eureka-math.org

Name _____   Date _____

1. Write a repeated addition equation to match the picture. Then, group the addends into pairs to show a more efficient way to add.

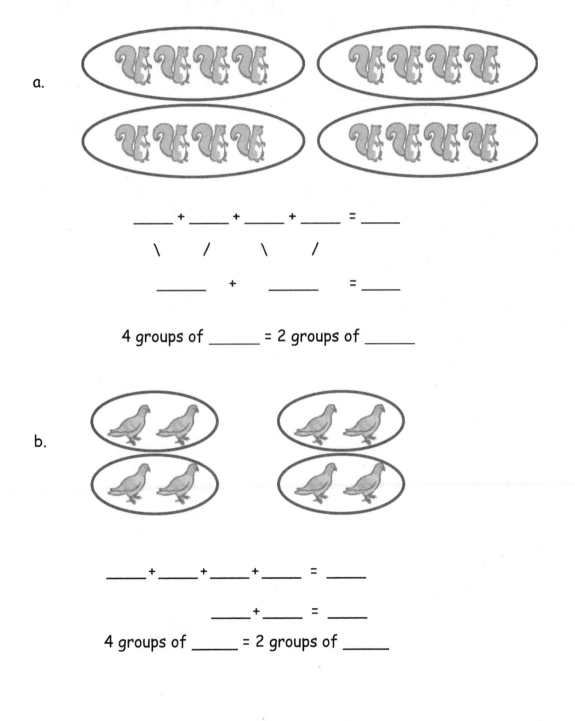

a.

_____ + _____ + _____ + _____ = _____

\\   /   \\   /

_____   +   _____   = _____

4 groups of _____ = 2 groups of _____

b.

_____ + _____ + _____ + _____ = _____

_____ + _____ = _____

4 groups of _____ = 2 groups of _____

    Lesson 3:   Use math drawings to represent equal groups, and relate to repeated   13
                            addition.

© 2018 Great Minds®. eureka-math.org

c.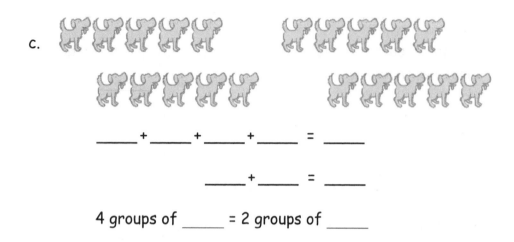

_____ + _____ + _____ + _____ = _____

_____ + _____ = _____

4 groups of _____ = 2 groups of _____

2. Write a repeated addition equation to match the picture. Then, group addends into pairs, and add to find the total.

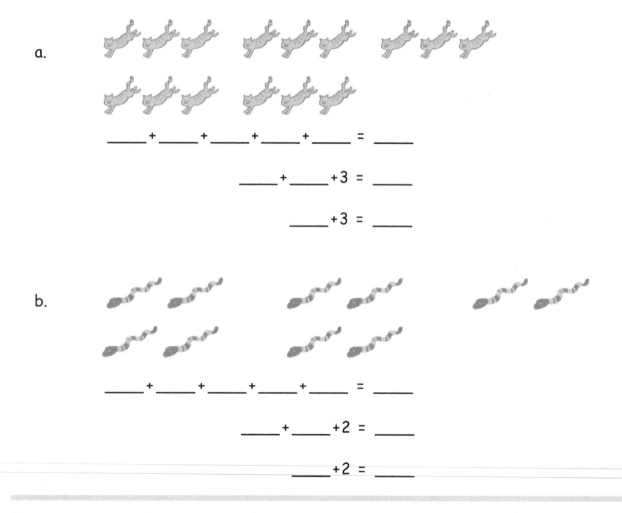

a.

_____ + _____ + _____ + _____ + _____ = _____

_____ + _____ + 3 = _____

_____ + 3 = _____

b.

_____ + _____ + _____ + _____ + _____ = _____

_____ + _____ + 2 = _____

_____ + 2 = _____

Lesson 3:   Use math drawings to represent equal groups, and relate to repeated addition.

1. Write a repeated addition equation to find the total of each tape diagram.

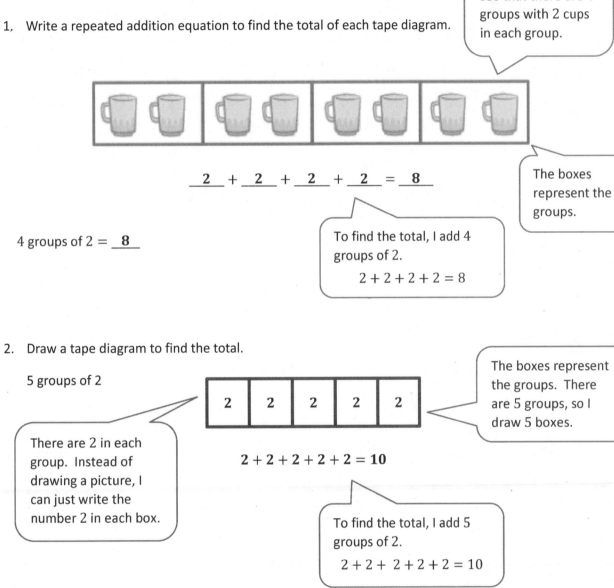

This tape diagram drawing helps me see that there are 4 groups with 2 cups in each group.

__2__ + __2__ + __2__ + __2__ = __8__

The boxes represent the groups.

4 groups of 2 = __8__

To find the total, I add 4 groups of 2.

$2 + 2 + 2 + 2 = 8$

2. Draw a tape diagram to find the total.

5 groups of 2

| 2 | 2 | 2 | 2 | 2 |

The boxes represent the groups. There are 5 groups, so I draw 5 boxes.

There are 2 in each group. Instead of drawing a picture, I can just write the number 2 in each box.

$2 + 2 + 2 + 2 + 2 = 10$

To find the total, I add 5 groups of 2.

$2 + 2 + 2 + 2 + 2 = 10$

Lesson 4: Represent equal groups with tape diagrams, and relate to repeated addition.

15

EUREKA MATH®

Name _____ Date _____

1. Write a repeated addition equation to find the total of each tape diagram.

    a.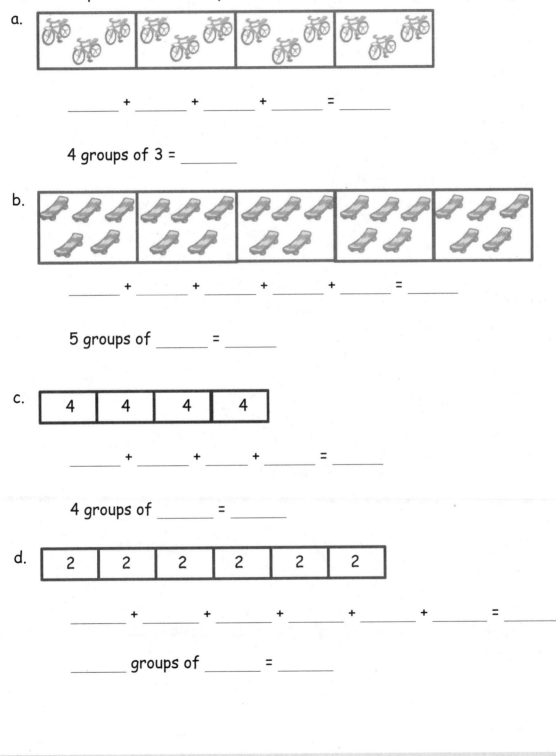

    _____ + _____ + _____ + _____ = _____

    4 groups of 3 = _____

    b.

    _____ + _____ + _____ + _____ + _____ = _____

    5 groups of _____ = _____

    c.

4	4	4	4

    _____ + _____ + _____ + _____ = _____

    4 groups of _____ = _____

    d.

2	2	2	2	2	2

    _____ + _____ + _____ + _____ + _____ + _____ = _____

    _____ groups of _____ = _____

Lesson 4:   Represent equal groups with tape diagrams, and relate to repeated addition.

17

© 2018 Great Minds®. eureka-math.org

2. Draw a tape diagram to find the total.

a. 5 + 5 + 5 + 5 = _____

b. 4 + 4 + 4 + 4 + 4 = _____

c. 4 groups of 2

d. 5 groups of 3

e.

Lesson 4: Represent equal groups with tape diagrams, and relate to repeated addition.

© 2018 Great Minds®. eureka-math.org

1. Circle groups of two. Redraw the groups of two as rows and then as columns.

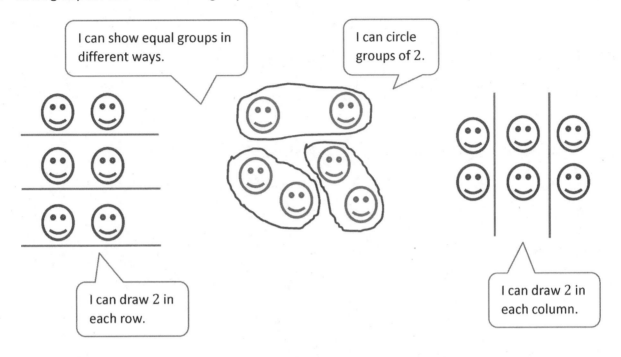

I can show equal groups in different ways.

I can circle groups of 2.

I can draw 2 in each row.

I can draw 2 in each column.

2. Count the objects in the array from left to right by rows and top to bottom by columns. As you count, circle the rows and then the columns.

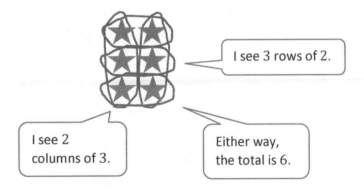

I see 3 rows of 2.

I see 2 columns of 3.

Either way, the total is 6.

 EUREKA MATH

Lesson 5: Compose arrays from rows and columns, and count to find the total using objects.

19

© 2018 Great Minds®. eureka-math.org

Name _____ Date _____

1. Circle groups of five. Then, draw the clouds into two equal rows.

_____

_____

2. Circle groups of four. Redraw the groups of four as rows and then as columns.

_____

_____

_____

3. Circle groups of four. Redraw the groups of four as rows and then as columns.

_____

_____

_____

Lesson 5: Compose arrays from rows and columns, and count to find the total using objects.

21

© 2018 Great Minds®. eureka-math.org

4. Count the objects in the arrays from left to right by rows and by columns. As you count, circle the rows and then the columns.

a.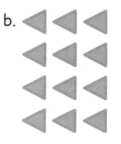

b.

5. Redraw the smiley faces and triangles in Problem 4 as columns of three.

6. Draw an array with 20 triangles.

7. Show a different array with 20 triangles.

Lesson 5:   Compose arrays from rows and columns, and count to find the total using objects.

© 2018 Great Minds®. eureka-math.org

Use the array of shaded triangles to answer the questions below.

a. **3** rows of **4** = 12

b. **4** columns of **3** = 12

c. **4** + **4** + **4** = **12**

d. Add 1 more row. How many triangles are there now? **16**

> When another row or column is added so is another group, or unit. I just think $12 + 4 = 16$.

e. Remove 1 column from the new array you made. How many triangles are there now? **12**

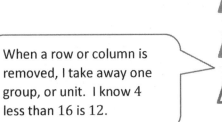

> When a row or column is removed, I take away one group, or unit. I know 4 less than 16 is 12.

Name _____   Date _____

1. Complete each missing part describing each array.

   Circle rows.   Circle columns.

   a.   b.

   3 rows of _____ = _____   4 columns of _____ = _____

   _____ + _____ + _____ = _____   _____ + _____ + _____ + _____ = _____

   Circle rows.   Circle columns.

   c.   d.

   5 rows of _____ = _____   3 columns of _____ = _____

   ___ + ___ + ___ + ___ + ___ = ___   ___ + ___ + ___ = ___

2. Use the array of smiley faces to answer the questions below.

    a. _____ rows of _____ = _____

    b. _____ columns of _____ = _____

    c. _____ + _____ + _____ = _____

    d. Add 1 more row. How many smiley faces are there now? _____

    e. Add 1 more column to the new array you made in 2(d). How many smiley faces are there now? _____

3. Use the array of squares to answer the questions below.

    a. _____ + _____ + _____ + _____ = _____

    b. _____ rows of _____ = _____

    c. _____ columns of _____ = _____

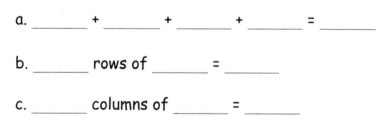

    d. Remove 1 row. How many squares are there now? _____

    e. Remove 1 column from the new array you made in 3(d). How many squares are there now? _____

Lesson 6:   Decompose arrays into rows and columns, and relate to repeated addition.

© 2018 Great Minds®. eureka-math.org

EUREKA
MATH

1. Draw an array with X's that has 3 columns of 4. Draw vertical lines to separate the columns. Fill in the blanks.

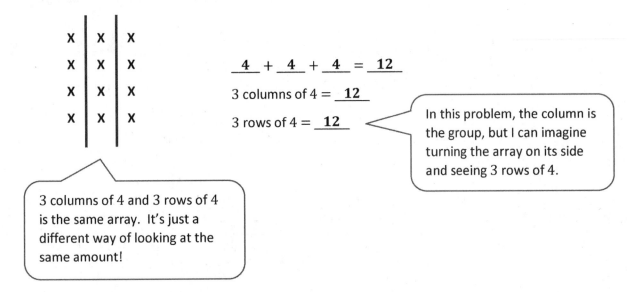

$\underline{\ 4\ } + \underline{\ 4\ } + \underline{\ 4\ } = \underline{\ 12\ }$

3 columns of 4 = __12__

3 rows of 4 = __12__

In this problem, the column is the group, but I can imagine turning the array on its side and seeing 3 rows of 4.

3 columns of 4 and 3 rows of 4 is the same array. It's just a different way of looking at the same amount!

2. Draw an array of X's with 1 more column of 4 than the array shown above. Write a repeated addition equation to find the total number of X's.

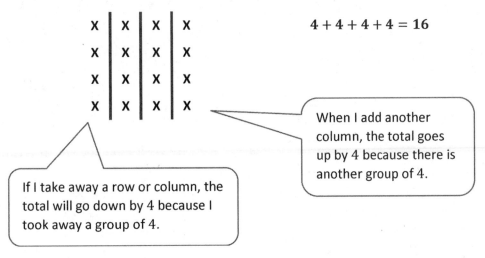

$4 + 4 + 4 + 4 = 16$

When I add another column, the total goes up by 4 because there is another group of 4.

If I take away a row or column, the total will go down by 4 because I took away a group of 4.

© 2018 Great Minds®. eureka-math.org

Name _____   Date _____

1. a   One row of an array is drawn below. Complete the array with X's to make 4 rows
        of 5. Draw horizontal lines to separate the rows.

        <u>X X X X X</u>

    b. Draw an array with X's that has 4 columns of 5. Draw vertical lines to separate
        the columns. Fill in the blanks.

        _____ + _____ + _____ + _____ = _____

                        4 rows of 5 = _____

                        4 columns of 5 = _____

2. a. Draw an array of X's with 3 columns of 4.

    b. Draw an array of X's with 3 rows of 4. Fill in the blanks below.

        _____ + _____ + _____ = _____

                        3 columns of 4 = _____

                        3 rows of 4 = _____

In the following problems, separate the rows or columns with horizontal or vertical lines.

3. Draw an array of X's with 3 rows of 3.

_____ + _____ + _____ = _____

3 rows of 3 = _____

4. Draw an array of X's with 2 more rows of 3 than the array in Problem 3. Write a repeated addition equation to find the total number of X's.

5. Draw an array of X's with 1 less column than the array in Problem 4. Write a repeated addition equation to find the total number of X's.

Lesson 7: Represent arrays and distinguish rows and columns using math drawings.

© 2018 Great Minds®. eureka-math.org

1. Create an array with the squares.

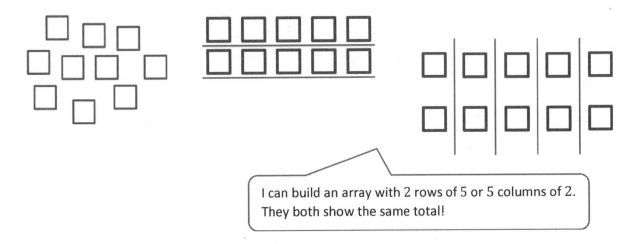

I can build an array with 2 rows of 5 or 5 columns of 2. They both show the same total!

2. Use the array of squares to answer the questions below.

Since there are 3 addends, I know this repeated addition equation relates to the columns.

a. There are __3__ squares in one row.

b. There are __4__ squares in one column.

c. __4__ + __4__ + __4__ = __12__

d. 3 columns of __4__ = __4__ rows of __3__ = __12__ total.

3. Draw a tape diagram to match your repeated addition equation and array.

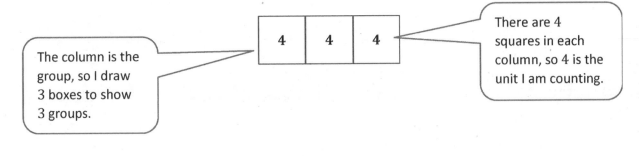

The column is the group, so I draw 3 boxes to show 3 groups.

| 4 | 4 | 4 |

There are 4 squares in each column, so 4 is the unit I am counting.

Name _____   Date _____

1. Create an array with the squares.

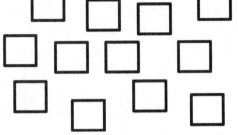

_____

_____

2. Create an array with the squares from the set above.

3. Use the array of squares to answer the questions below.

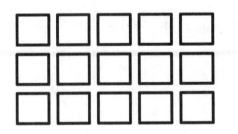

a. There are _____ squares in each row.

b. _____ + _____ + _____ = _____

c. There are _____ squares in each column.

d. _____ + _____ + _____ + _____ + _____ = _____

4. Use the array of squares to answer the questions below.

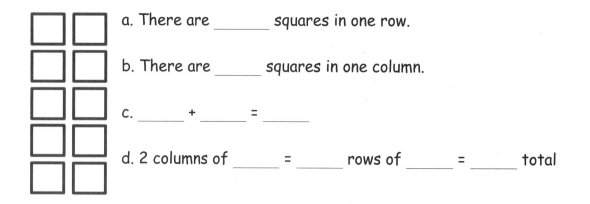

    a. There are _____ squares in one row.

    b. There are _____ squares in one column.

    c. _____ + _____ = _____

    d. 2 columns of _____ = _____ rows of _____ = _____ total

5. a Draw an array with 15 squares that has 3 squares in each column.

    b. Write a repeated addition equation to match the array.

6. a. Draw an array with 20 squares that has 5 squares in each column.

    b. Write a repeated addition equation to match the array.

    c. Draw a tape diagram to match your repeated addition equation and array.

Lesson 8:   Create arrays using square tiles with gaps.

© 2018 Great Minds®. eureka-math.org

1. Draw an array for each word problem. Write a repeated addition equation to match each array.

    Jason collected some stones. He put them in 5 rows with 3 stones in each row. How many stones did Jason have altogether?

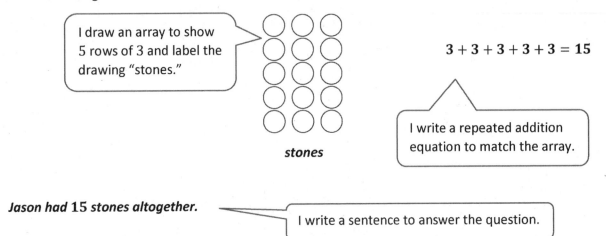

    I draw an array to show 5 rows of 3 and label the drawing "stones."

    stones

    $3 + 3 + 3 + 3 + 3 = 15$

    I write a repeated addition equation to match the array.

    Jason had 15 stones altogether.

    I write a sentence to answer the question.

2. Draw a tape diagram for each word problem. Write a repeated addition equation to match each tape diagram.

    Each of Maria's 4 friends has 5 markers. How many markers do Maria's friends have in all?

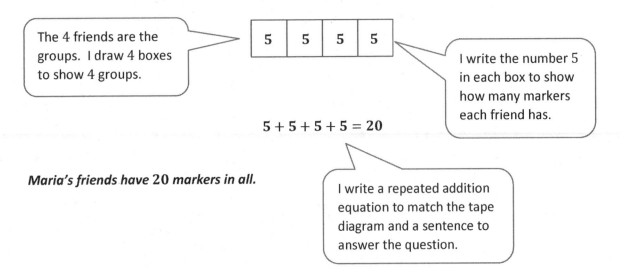

    The 4 friends are the groups. I draw 4 boxes to show 4 groups.

    | 5 | 5 | 5 | 5 |

    I write the number 5 in each box to show how many markers each friend has.

    $5 + 5 + 5 + 5 = 20$

    Maria's friends have 20 markers in all.

    I write a repeated addition equation to match the tape diagram and a sentence to answer the question.

Name _____   Date _____

Draw an array for each word problem. Write a repeated addition equation to match each array.

1. Melody stacked her blocks in 3 columns of 4. How many blocks did Melody stack in all?

2. Marty arranged the desks in the classroom into 5 equal rows. There were 5 desks in each row. How many desks were arranged?

3. The baker made 5 trays of muffins. Each tray holds 4 muffins. How many muffins did the baker make?

4. The library books were on the shelf in 4 stacks of 4. How many books were
    on the shelf?

Draw a tape diagram for each word problem. Write a repeated addition equation
to match each tape diagram.

5. Mary placed stickers in columns of 4. She made 5 columns. How many stickers
    did she use?

6. Jayden put his baseball cards into 5 columns of 3 in his book. How many cards did
    Jayden put in his book?

Draw a tape diagram and an array. Then, write a repeated addition equation to match.

7. The game William bought came with 3 bags of marbles. Each bag had 3 marbles
    inside. How many total marbles came with the game?

Lesson 9:   Solve word problems involving addition of equal groups in rows and columns.

1. Use your square tiles to construct the following rectangles with no gaps or overlaps. Write a repeated addition equation to match each construction.

Construct a rectangle with 2 rows of 3 tiles.   Construct a rectangle with 2 columns of 3 tiles.

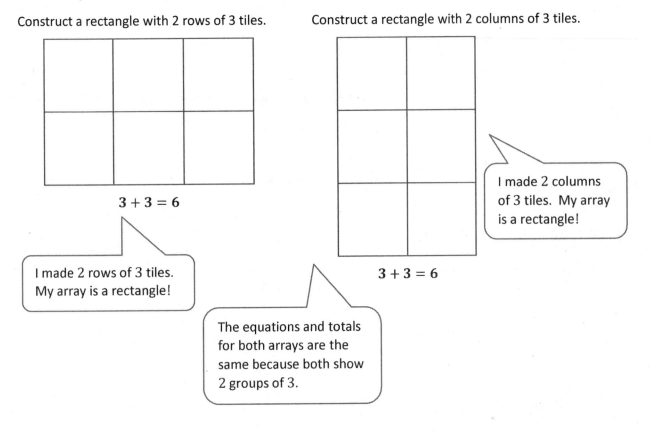

$$3 + 3 = 6$$

$$3 + 3 = 6$$

I made 2 columns of 3 tiles. My array is a rectangle!

I made 2 rows of 3 tiles. My array is a rectangle!

The equations and totals for both arrays are the same because both show 2 groups of 3.

2. Construct a rectangle of 4 tiles that has equal rows and columns. Write a repeated addition equation to match.

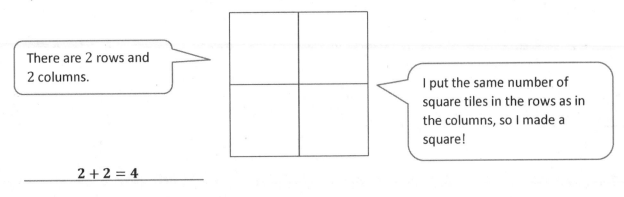

There are 2 rows and 2 columns.

I put the same number of square tiles in the rows as in the columns, so I made a square!

$$2 + 2 = 4$$

EUREKA
MATH®

Name _____   Date _____

Cut out the square tiles below, and construct the following arrays with no gaps
or overlaps. On the line, write a repeated addition equation to match each construction
on the line.

1. a. Construct a rectangle with   b. Construct a rectangle with
       2 rows of 4 tiles.   2 columns of 4 tiles.

       _____   _____

2. a. Construct a rectangle with   b. Construct a rectangle with
       3 rows of 2 tiles.   3 columns of 2 tiles.

       _____   _____

3. a. Construct a rectangle   b. Construct a rectangle
       using 10 tiles.   using 12 tiles.

       _____   _____

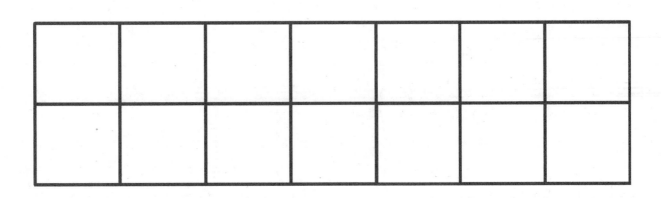

 EUREKA MATH®

Lesson 10:   Use square tiles to compose a rectangle, and relate to the array model.

41

© 2018 Great Minds®. eureka-math.org

4. a. What shape is the array pictured below? _____

b. In the space below, redraw the above shape with one more column.

c. What shape is the array now? _____

d. Draw a different array of tiles that is the same shape as 4(c).

1. Construct an array with 20 square tiles.

    Write a repeated addition equation to match the array.

$$5 + 5 + 5 + 5 = 20$$

				5

5
5
5
5

Rearrange the 20 square tiles into a different array.

10
10

I can make an array with 4 rows of 5 tiles and write a repeated addition equation to match. It's easy to skip-count by 5's.

Write a repeated addition equation to match the new array.

$$10 + 10 = 20$$

I can rearrange the tiles to make another array with 2 rows of 10 tiles. I can use my doubles facts to find the total: $10 + 10 = 20$.

2. Construct 2 arrays with 16 square tiles.

   2 rows of **8** = **16**

   If I turn 2 rows of 8 so they're standing up, I will have 8 rows of 2. I know that $8 + 8$ equals $2 + 2 + 2 + 2 + 2 + 2 + 2 + 2$.

   2 rows of **8** = 8 rows of **2**

Name _____   Date _____

1. a. Construct an array with 9 square tiles.

    b. Write a repeated addition equation to match the array.

    _____

2. a. Construct an array with 10 square tiles.

    b. Write a repeated addition equation to match the array.

    _____

    c   Rearrange the 10 square tiles into a different array.

    d. Write a repeated addition equation to match the new array.

    _____

Cut out each square tile. Use the tiles to construct the arrays in Problems 1-4.

3. a. Construct an array with 12 square tiles.

    b. Write a repeated addition equation to match the array.

    _____

    c. Rearrange the 12 square tiles into a different array.

    d   Write a repeated addition equation to match the new array.

    _____

4. Construct 2 arrays with 14 square tiles.

    a. 2 rows of _____ = _____

    b. 2 rows of _____ = 7 rows of _____

Lesson 11:   Use square tiles to compose a rectangle, and relate to the array model.

EUREKA
MATH

1. Trace a square tile to make an array with 3 columns of 4.

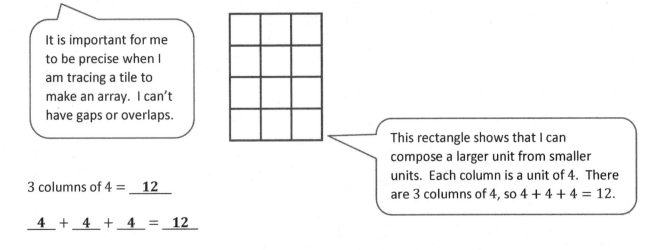

It is important for me to be precise when I am tracing a tile to make an array. I can't have gaps or overlaps.

This rectangle shows that I can compose a larger unit from smaller units. Each column is a unit of 4. There are 3 columns of 4, so $4 + 4 + 4 = 12$.

3 columns of 4 = __12__

__4__ + __4__ + __4__ = __12__

2. Complete the following array without gaps or overlaps. The first tile has been drawn for you.

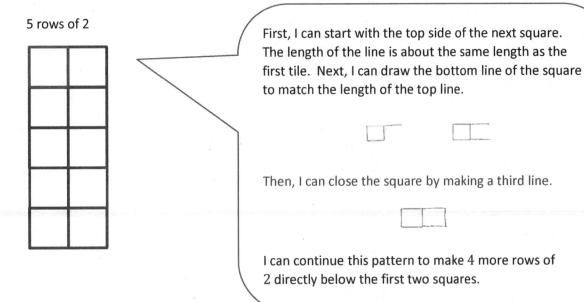

5 rows of 2

First, I can start with the top side of the next square. The length of the line is about the same length as the first tile. Next, I can draw the bottom line of the square to match the length of the top line.

Then, I can close the square by making a third line.

I can continue this pattern to make 4 more rows of 2 directly below the first two squares.

Lesson 12: Use math drawings to compose a rectangle with square tiles.

47

© 2018 Great Minds®. eureka-math.org

Name _____   Date _____

1. Cut out and trace the square tile to draw an array with 2 rows of 4.

   ┌─────────┐
   │ Cut out │
   │ and trace. │
   └─────────┘

   2 rows of 4 = _____

   _____ + _____ = _____

2. Trace the square tile to make an array with 3 columns of 5.

   3 columns of 5 = _____

   _____ + _____ + _____ = _____

3. Complete the following arrays without gaps or overlaps. The first tile has been drawn for you.

   a. 4 rows of 5

   b. 5 columns of 2

   c. 4 columns of 3

Lesson 12:   Use math drawings to compose a rectangle with square tiles.

1. Step 1: Construct a rectangle with 5 columns of 3.

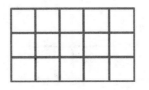

Step 2: Separate 3 columns of 3.

I decompose 5 columns of 3 into 2 smaller rectangles, or parts. 3 columns of 3 and 2 columns of 3 make 5 columns of 3.

Step 3: Write a number bond to show the whole and two parts. Write a repeated addition sentence to match each part of the number bond.

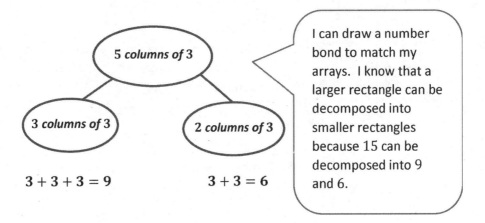

5 *columns of* 3

3 *columns of* 3   2 *columns of* 3

$3 + 3 + 3 = 9$ $3 + 3 = 6$

I can draw a number bond to match my arrays. I know that a larger rectangle can be decomposed into smaller rectangles because 15 can be decomposed into 9 and 6.

EUREKA
MATH

2. Use 16 square tiles to construct a rectangle.

   a. __4__ rows of __4__ = __16__

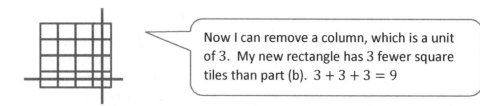

   > I can remove a row, which is a unit of 4, so my new rectangle has 12 square tiles. $4 + 4 + 4 = 12$

   b. Remove 1 row. How many square tiles are there now? __12__

   c. Remove 1 column from the new rectangle you made in part (b). How many square tiles are there now? __9__

   > Now I can remove a column, which is a unit of 3. My new rectangle has 3 fewer square tiles than part (b). $3 + 3 + 3 = 9$

     Lesson 13:   Use square tiles to decompose a rectangle.

Name _____ Date _____

Cut out and use your square tiles to complete the steps for each problem.

Problem 1

    Step 1: Construct a rectangle with 5 rows of 2.

    Step 2: Separate 2 rows of 2.

    Step 3: Write a number bond to show the whole and two parts. Write a repeated addition sentence to match each part of your number bond.

Problem 2

    Step 1: Construct a rectangle with 4 columns of 3.

    Step 2: Separate 2 columns of 3.

    Step 3: Write a number bond to show the whole and two parts. Write a repeated addition sentence to match each part of your number bond.

3. Use 9 square tiles to construct a rectangle with 3 rows.

a. _____ rows of _____ = _____

b. Remove 1 row. How many squares are there now? _____

c. Remove 1 column from the new rectangle you made in 3(b). How many squares are there now? _____

4. Use 14 square tiles to construct a rectangle.

a. _____ rows of _____ = _____

b. Remove 1 row. How many squares are there now? _____

c. Remove 1 column from the new rectangle you made in 4(b). How many squares are there now? _____

EUREKA
MATH

square tiles

EUREKA
MATH®

1. Imagine that you have just cut this rectangle into rows.

    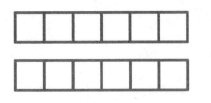

    a. What do you see? Draw a picture.

    > I can decompose the same rectangle into rows and columns. I can see 2 rows of 6.

    How many squares are in each row? __6__

    b. Imagine that you have just cut this rectangle into columns. What do you see? Draw a picture.

    How many squares are in each column? __2__

    > I can also see 6 columns of 2.

2. Create another rectangle using the same number of squares.

    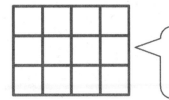

    > I can make another rectangle with the same 12 squares. I can rearrange 2 columns of 2 as 1 row of 4. Now, my rectangle has 3 rows of 4.

    How many squares are in each row? __4__

    How many squares are in each column? __3__

Lesson 14:   Use scissors to partition a rectangle into same-size squares, and compose arrays with the squares.

57

© 2018 Great Minds®. eureka-math.org

Name _____   Date _____

1. Imagine that you have just cut this rectangle into rows.

    a. What do you see? Draw a picture.

        How many squares are in each row? _____

    b. Imagine that you have just cut this rectangle into columns. What do you see? Draw a picture.

        How many squares are in each column? _____

2. Create another rectangle using the same number of squares.

    How many squares are in each row? _____
    How many squares are in each column? _____

Lesson 14:   Use scissors to partition a rectangle into same-size squares, and compose arrays
                with the squares.

© 2018 Great Minds®. eureka-math.org

59

3. Imagine that you have just cut this rectangle into rows.

    a. What do you see? Draw a picture.

    How many squares are in each row? _____

    b. Imagine that you have just cut this rectangle into columns. What do you see?
        Draw a picture.

    How many squares are in each column? _____

4. Create another rectangle using the same number of squares.

    How many squares are in each row? _____
    How many squares are in each column? _____

Lesson 14:   Use scissors to partition a rectangle into same-size squares, and compose arrays
               with the squares.

© 2018 Great Minds®. eureka-math.org

1. Shade in an array with 5 columns of 4.

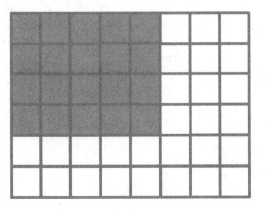

I can shade 1 column of 4 and then 4 more columns of 4. I can say that each column has a group, or unit, of 4.

Write a repeated addition equation for the array.

$$4 + 4 + 4 + 4 + 4 = 20$$

I see 5 columns of 4, or 5 fours. I can use doubles to add. $8 + 8 + 4 = 20$. I have shaded 20 squares altogether.

2. Draw one more row and then two more columns to make a new array.

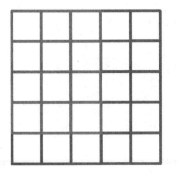

First, I can draw another row of 3. Now there are 5 rows of 3. Then I can draw 2 more columns. That makes 5 columns of 5 altogether.

Write a repeated addition equation for the new array.

$$5 + 5 + 5 + 5 + 5 = 25$$

I see 5 columns of 5, or 5 fives. I can skip-count by 5's. There are 25 squares in all.

Lesson 15:   Use math drawings to partition a rectangle with square tiles, and relate to repeated addition.

61

© 2018 Great Minds®. eureka-math.org

Name _____   Date _____

1. Shade in an array with 3 rows of 2.

Write a repeated addition
equation for the array.

_____

2. Shade in an array with 2 rows of 4.

Write a repeated addition
equation for the array.

_____

3. Shade in an array with 4 columns of 5.

Write a repeated addition
equation for the array.

_____

          Lesson 15:   Use math drawings to partition a rectangle with square tiles, and   **63**
                             relate to repeated addition.

© 2018 Great Minds®. eureka-math.org

4. Draw one more column of 2 to make a new array.

Write a repeated addition
equation for the new array.

_____

5. Draw one more row of 3 and then one more column to make a new array.

Write a repeated addition
equation for the new array.

_____

6. Draw one more row and then two more columns to make a new array.

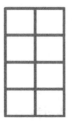

Write a repeated addition
equation for the new array.

_____

Lesson 15:   Use math drawings to partition a rectangle with square tiles, and
                              relate to repeated addition.

1. Shade to create a copy of the design on the empty grid.

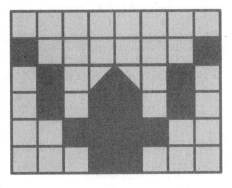   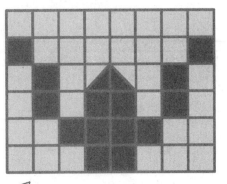

I can use square tiles to put together and break apart rectangles. Look, I see that some squares are only half-shaded to make triangles! When I make designs, I have to pay close attention to the rows and columns so that I shade in the correct squares.

2. Use colored pencils to create a design in the bolded square section. Create a tessellation by repeating the design throughout.

The core unit that I am repeating has 3 rows and 3 columns. I can create the same design again by shading in the same pattern. I know that this pattern could go on and on if I kept repeating it.

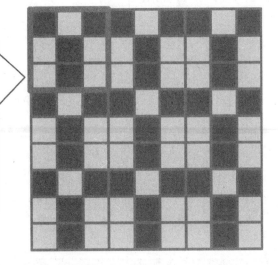

Name _____   Date _____

1. Shade to create a copy of the design on the empty grid.

   a.

   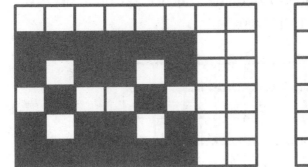

   b.

   c.

   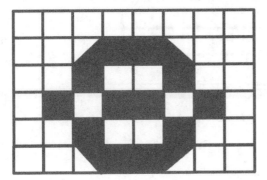

Lesson 16:   Use grid paper to create designs to develop spatial structuring.

67

2. Create two different designs.

3. Use colored pencils to create a design in the bolded square section. Create a tessellation by repeating the design throughout.

Lesson 16:   Use grid paper to create designs to develop spatial structuring.

1. Draw to double the group you see. Complete the sentences, and write an addition equation.

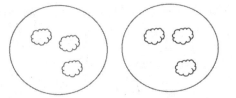

There are __3__ clouds in each group.

__3__ + __3__ = __6__

I know that when both addends are the same, I have doubles. $1 + 1 = 2$, $2 + 2 = 4$, $3 + 3 = 6$, and so on. Doubling a number always makes an even number even when there are 3 objects in each group.

2. Draw an array for the set below. Complete the sentences.

2 rows of 5

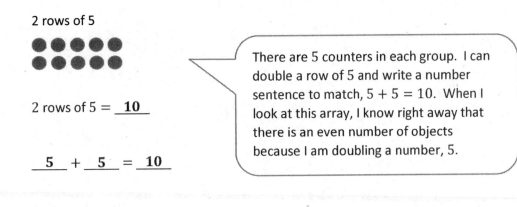

2 rows of 5 = __10__

There are 5 counters in each group. I can double a row of 5 and write a number sentence to match, $5 + 5 = 10$. When I look at this array, I know right away that there is an even number of objects because I am doubling a number, 5.

__5__ + __5__ = __10__

5 doubled is __10__ .

Name _____   Date _____

1. Draw to double the group you see. Complete the sentences, and write
   an addition equation.

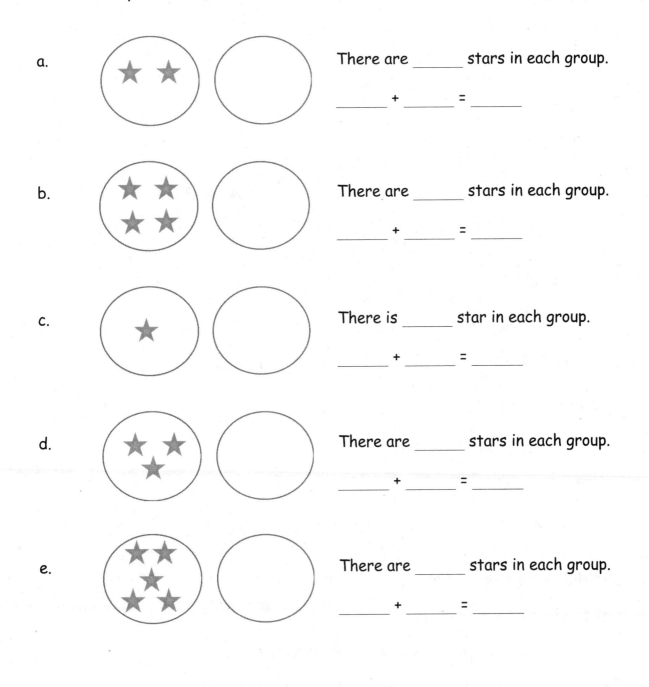

a.   There are _____ stars in each group.

                  _____ + _____ = _____

b.   There are _____ stars in each group.

                  _____ + _____ = _____

c.   There is _____ star in each group.

                  _____ + _____ = _____

d.   There are _____ stars in each group.

                  _____ + _____ = _____

e.   There are _____ stars in each group.

                  _____ + _____ = _____

© 2018 Great Minds®. eureka-math.org

2. Draw an array for each set. Complete the sentences. The first one has been drawn for you.

a. **2 rows of 6**

2 rows of 6 = _____

_____ + _____ = _____

6 doubled is _____.

b. **2 rows of 7**

2 rows of 7 = _____

_____ + _____ = _____

7 doubled is _____.

c. **2 rows of 8**

_____ rows of _____ = _____

_____ + 8 = _____

8 doubled is _____.

d. **2 rows of 9**

2 rows of 9 = _____

_____ + _____ = _____

9 doubled is _____.

e. **2 rows of 10**

_____ rows of _____ = _____

10 + _____ = _____

10 doubled is _____.

3. List the totals from Problem 1. _____

List the totals from Problem 2. _____

Are the numbers you have listed even or not even? _____

Explain in what ways the numbers are the same and different.

Lesson 17: Relate doubles to even numbers, and write number sentences to express the sums.

© 2018 Great Minds®. eureka-math.org

EUREKA MATH

1. Pair the objects, and count by twos to decide if the number of objects is even.

(Even)/Not Even

There are 10 stars. The number of objects is even because when I pair them, there are no stars left over.

There are **5** twos. There are **0** twos left over.

Count by twos to find the total.

2 , **4** , **6** , **8** , **10**

10 is even because I can say 10 when counting by twos.

2. Draw to continue the pattern of the pairs in the space below until you have drawn 10 pairs.

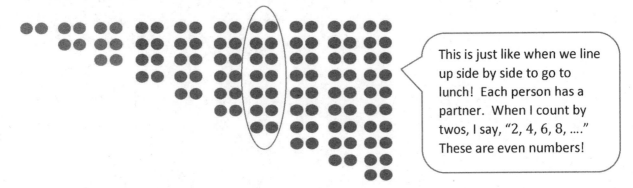

This is just like when we line up side by side to go to lunch! Each person has a partner. When I count by twos, I say, "2, 4, 6, 8, …." These are even numbers!

3. Write the number of dots in each array in Problem 2 in order from least to greatest.

    2, 4, 6, 8, 10, 12, 14, 16, 18, 20

4. Circle the array in Problem 2 that has 2 columns of 7.

    I can make 2 columns of 7, and $7 + 7 = 14$. Even if one of the numbers I'm adding isn't even, when I double it, I get an even number.

EUREKA MATH®

Name _____   Date _____

1. Pair the objects to decide if the number of objects is even.

Even/Not Even

_____

Even/Not Even

_____

Even/Not Even

2. Draw to continue the pattern of the pairs in the spaces below until you have drawn
    zero pairs.

3. Write the number of hearts in each array in Problem 2 in order from greatest to least.

4. Circle the array in Problem 2 that has 2 columns of 6.

5. Box the array in Problem 2 that has 2 columns of 8.

6. Redraw the set of stars as columns of two or 2 equal rows.

There are _____ stars.

Is _____ an even number? _____

7. Circle groups of two. Count by twos to see if the number of objects is even.

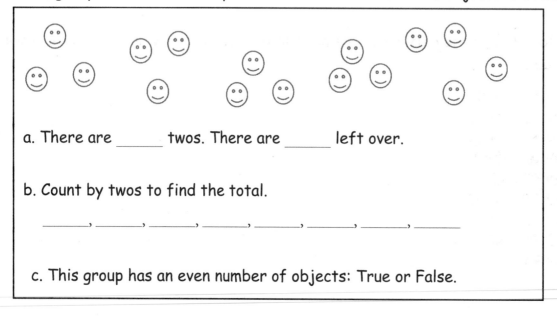

a. There are _____ twos. There are _____ left over.

b. Count by twos to find the total.

_____, _____, _____, _____, _____, _____, _____, _____

c. This group has an even number of objects: True or False.

Lesson 18:   Pair objects and skip-count to relate to even numbers.

EUREKA MATH

1. Skip-count the columns in the array.   The first one has been done for you.

○ ○ ○ ○ ○ ○
○ ○ ○ ○ ○ ○

__2__ __4__ __6__ __8__ __10__ __12__

> I can skip-count by 2's using the columns of the array. If I keep adding columns of 2 to this pattern, I can say, "..., 14, 16, 18, 20." There's a pattern in the ones place! 0, 2, 4, 6, 8.

2. Solve.

$1 + 1 =$ __2__ $4 + 4 =$ __8__

$2 + 2 =$ __4__ $5 + 5 =$ __10__

$3 + 3 =$ __6__ $6 + 6 =$ __12__

> When I find doubles, I see a pattern in the answers; they are skip-counting by 2's.

3. Write to identify the **bold** numbers as *even* or *odd*.

$24 + 1 = 25$	$24 - 1 = 23$
even + 1 = *odd*	*even* − 1 = *odd*

> When I add 1 to or subtract 1 from an even number, the new number is always odd!

4. Is the **bold** number even or odd? Circle the answer, and explain how you know.

39 even/odd	**Explanation:** This number does not have 0, 2, 4, 6, or 8 in the ones place. I know that 40 is even, so 40 − 1 has to be odd.

Lesson 19:   Investigate the pattern of even numbers: 0, 2, 4, 6, and 8 in the ones place, and relate to odd numbers.   **77**

© 2018 Great Minds®. eureka-math.org

Name _____   Date _____

1. Skip-count the columns in the array. The first one has been done for you.

○ ○ ○ ○ ○ ○ ○ ○ ○ ○
○ ○ ○ ○ ○ ○ ○ ○ ○ ○

2
___ ___ ___ ___ ___ ___ ___ ___ ___ ___

2. a. Solve.

$1 + 1 =$ _____ $6 + 6 =$ _____

$2 + 2 =$ _____ $7 + 7 =$ _____

$3 + 3 =$ _____ $8 + 8 =$ _____

$4 + 4 =$ _____ $9 + 9 =$ _____

$5 + 5 =$ _____ $10 + 10 =$ _____

b. How is the array in Problem 1 related to the answers in Problem 2(a)?

_____

_____

_____

3. Fill in the missing even numbers on the number path.

18, 20, _____, _____, 26, _____, 30, _____, 34, _____, 38, 40, _____, _____

EUREKA
MATH   Lesson 19:   Investigate the pattern of even numbers: 0, 2, 4, 6, and 8 in the ones   79
place, and relate to odd numbers.

© 2018 Great Minds®. eureka-math.org

4. Fill in the missing odd numbers on the number path.

0, _____, 2, _____, 4, _____, 6, _____, 8, _____, 10, _____, 12, _____, 14

5. Write to identify the **bold** numbers as even or odd. The first one has been done for you.

a.	b.	c.
4 + 1 = 5 _even_ + 1 = _odd_	**13 + 1 = 14** _____ + 1 = _____	**20 + 1 = 21** _____ + 1 = _____
d.	e.	f.
8 – 1 = 7 _____ – 1 = _____	**16 – 1 = 15** _____ – 1 = _____	**30 – 1 = 29** _____ – 1 = _____

6. Are the **bold** numbers even or odd? Circle the answer, and explain how you know.

a. **21** even/odd	Explanation:
b. **34** even/odd	Explanation:

     Lesson 19:   Investigate the pattern of even numbers: 0, 2, 4, 6, and 8 in the ones place, and relate to odd numbers.

© 2018 Great Minds®. eureka-math.org

1. Use the objects to create an array.

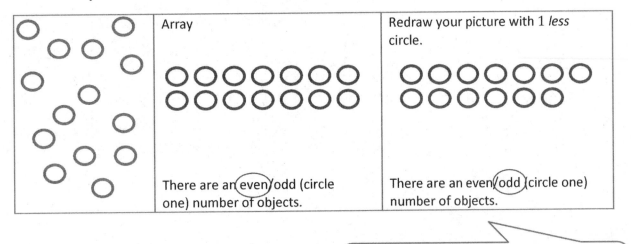

	Array	Redraw your picture with 1 *less* circle.
(scattered circles)	○○○○○○○○ ○○○○○○○○	○○○○○○○○ ○○○○○○○
	There are an (even)/odd (circle one) number of objects.	There are an even/(odd) (circle one) number of objects.

If I draw the array with 1 less circle, there are an odd number of objects. Now, I don't see 2 equal groups of 7.

2. Solve. Tell if each number is odd (0) or even (E).

$$11 \ + \ 13 \ = \ \underline{\ 24 \ }$$

$$\underline{\ O\ } \ + \ \underline{\ O \ } \ = \ \underline{\ E \ }$$

I know that 11 and 13 are odd because they do not have 0, 2, 4, 6 , or 8 in the ones place. When I add two odd numbers, I get an even number.

3. Write two examples for each case; next to your answer, write if your answers are even or odd. Add an even number to an odd number.

$$\underline{12 + 7 = 19 \ \ odd} \qquad \underline{8 + 13 = 21 \ \ odd}$$

I know that when I add an even number and an odd number, the sum will be odd. I cannot make 2 equal groups with 21 tiles, and I can't count by twos to 21.

Lesson 20: Use rectangular arrays to investigate odd and even numbers.

81

Name _____   Date _____

1. Use the objects to create an array with 2 rows.

a.	Array with 2 rows	Redraw your picture with 1 *less* star.
	There are an even/odd (circle one) number of stars.	There are an even/odd (circle one) number of stars.
b.	Array with 2 rows	Redraw your picture with 1 *more* star.
	There are an even/odd (circle one) number of stars.	There are an even/odd (circle one) number of stars.
c.	Array with 2 rows	Redraw your picture with 1 *less* star.
	There are an even/odd (circle one) number of stars.	There are an even/odd (circle one) number of stars.

EUREKA MATH

Lesson 20:   Use rectangular arrays to investigate odd and even numbers.

© 2018 Great Minds®. eureka-math.org

83

2. Solve. Tell if each number is odd (O) or even (E) on the line below.

    a.   6   +   6   = _____   e.   7   +   8   = _____

        _____ + _____ = _____   _____ + _____ = _____

    b.   8   +   13   = _____   f.   9   +   11   = _____

        _____ + _____ = _____   _____ + _____ = _____

    c.   9   +   15   = _____   g.   7   +   14   = _____

        _____ + _____ = _____   _____ + _____ = _____

    d.   17   +   8   = _____   h.   9   +   9   = _____

        _____ + _____ = _____   _____ + _____ = _____

3. Write three number sentence examples to prove that each statement is correct.

Even + Even = Even	Even + Odd = Odd	Odd + Odd = Even

EUREKA
MATH

4. Write two examples for each case. Next to your answer, write if your answers are even or odd. The first one has been done for you.

a. Add an even number to an even number.

_____32 + 8 = 40 even_____   _____

b. Add an odd number to an even number.

_____   _____

c. Add an odd number to an odd number.

_____   _____

Grade 2
Module 7

1. Count and categorize each picture to complete the table with tally marks.

No Legs	2 Legs	4 Legs
I	III	III

I can count how many animals are in each category. I cross out each animal as I record it with a tally mark under the correct category.

2. Use the Animal Classification table to answer the following questions about the types of animals Ms. Lee's second-grade class found in the local zoo.

Animal Classification			
Birds	Fish	Mammals	Reptiles
6	5	11	3

I know that this question is asking me to find the total number of birds, fish, or reptiles in the table. It's not asking for the number of categories.

a. How many animals are birds, fish, or reptiles? __14__ $6 + 5 + 3 = 14$

b. How many more birds and mammals are there than fish and reptiles? __9__ $17 - 8 = 9$

c. How many animals were classified? __25__ $6 + 5 + 11 + 3 = 11 + 14 = 25$

d. If 5 more birds and 2 more reptiles were added to the table, how many fewer reptiles would there be than birds? __6__

   B $6 + 5 = 11$ $5 + \underline{6} = 11$

   R $3 + 2 = 5$

I can use addition or subtraction when I see the words *how many fewer*.

Lesson 1:   Sort and record data into a table using up to four categories; use category counts to solve word problems.

89

Name _____   Date _____

1. Count and categorize each picture to complete the table with tally marks.

No Legs	2 Legs	4 Legs

2. Count and categorize each picture to complete the table with numbers.

Fur	Feathers

Lesson 1:   Sort and record data into a table using up to four categories; use   91
                 category counts to solve word problems.

© 2018 Great Minds®. eureka-math.org

3. Use the Animal Habitats table to answer the following questions.

Animal Habitats		
Arctic	Forest	Grasslands
6	11	9

a. How many animals live in the arctic? ____

b. How many animals have habitats in the forest and grasslands? ____

c. How many fewer animals have arctic habitats than forest habitats? ____

d. How many more animals would need to be in the grasslands category to have the same number as the arctic and forest categories combined? ____

e. How many total animal habitats were used to create this table? ____

Lesson 1:   Sort and record data into a table using up to four categories; use
               category counts to solve word problems.

4. Use the Animal Classification table to answer the following questions about the class pets in West Chester Elementary School.

Animal Classification			
Birds	Fish	Mammals	Reptiles
7	15	18	9

a. How many animals are birds, fish, or reptiles? ____

b. How many more birds and mammals are there than fish and reptiles? ____

c. How many animals were classified? ____

d. If 3 more birds and 4 more reptiles were added to the table, how many fewer birds would there be than reptiles? ____

Lesson 1:   Sort and record data into a table using up to four categories; use category counts to solve word problems.

© 2018 Great Minds®. eureka-math.org

93

1. Use grid paper to create a picture graph below using data provided in the table. Then, answer the questions.

Central Park Zoo Animal Classification			
Birds	Fish	Mammals	Reptiles
6	5	11	3

Title: *Central Park Zoo Animal Classification*

a. How many more animals are mammals and fish than birds and reptiles? __7__

$$11 + 5 = 16 \qquad 6 + 3 = 9 \qquad 16 - 9 = 7$$

b. How many fewer animals are reptiles than mammals?
    __8__ $11 - 3 = 8$

> I use the graph to help me answer comparison questions like *how many more* or *how many fewer*.

Birds   Fish   Mammals   Reptiles

Legend: *Each* ◯ *stands for* **1 animal**

> I organize the data from the table in a vertical picture graph. I put the categories in the same order as they are in the table, so I don't get confused. I must remember to include a title and a legend.

Lesson 2:   Draw and label a picture graph to represent data with up to four categories.

© 2018 Great Minds®. eureka-math.org

2. Use the table below to create a picture graph in the space provided.

Animal Habitats		
Desert	Tundra	Grassland

I draw a circle in each box to represent each animal recorded by a tally mark in the table. Circles help me to draw efficiently, and the legend explains what they represent.

Title: ___Animal Habitats___

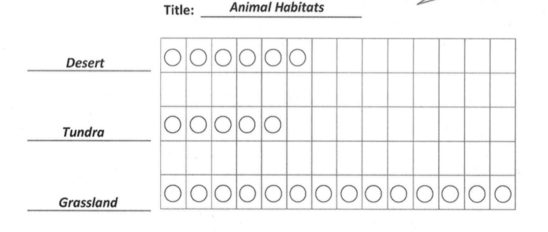

Desert

Tundra

Grassland

Legend: ___Each ◯ stands for 1 animal___

a. How many more animals live in the grassland than in the desert? __8__

$$14 - 6 = 8$$

b. How many fewer animals live in the tundra than in the grassland and desert combined? ___15___

$$14 + 6 = 20 \qquad 20 - 5 = 15$$

The first question asks *how many more*. I can figure out the answer by subtracting or by counting the extra circles in the picture graph for the grassland compared to the desert. There are 8 extra circles.

Draw and label a picture graph to represent data with up to four categories.

© 2018 Great Minds®. eureka-math.org

Name _____ Date _____

1. Use grid paper to create a picture graph below using data provided in the table. Then, answer the questions.

Favorite Mammals			
Tiger	Panda	Snow Leopard	Gorilla
8	11	7	12

Title: _____

a. How many more people chose gorilla as their favorite mammal than chose tiger? _____

b. How many more people chose tiger and gorilla as their favorite mammals than panda and snow leopard? _____

c. How many fewer people chose tiger as their favorite mammal than panda? _____

_____   _____   _____   _____

Legend: _____

d. Write and answer your own comparison question based on the data.

Question: _____

Answer: _____

Lesson 2:   Draw and label a picture graph to represent data with up to four categories.

© 2018 Great Minds®. eureka-math.org

97

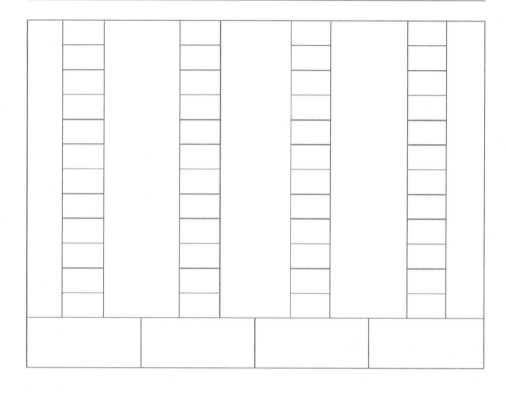

Legend: _____

Legend: _____

Lesson 2: Draw and label a picture graph to represent data with up to four categories.

2. Use the data of Mr. Clark's class vote to create a picture graph in the space provided.

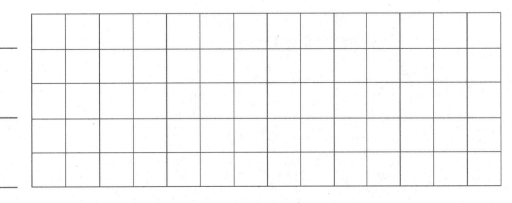

Favorite Birds		
Penguin	Flamingo	Peacock
⊮⊮I	⊮⊮	⊮⊮ ⊮⊮ IIII

Title: _____

Legend: _____

a. How many more students voted for peacocks than penguins? _____

b. How many fewer votes are for flamingos than penguins and peacocks? _____

c. Write and answer your own comparison question based on the data.

Question: _____

Answer: _____

Lesson 2:   Draw and label a picture graph to represent data with up to four categories.

© 2018 Great Minds®. eureka-math.org

99

Complete the bar graph below using data provided in the table.

Animal Habitats		
Desert	Arctic	Grassland

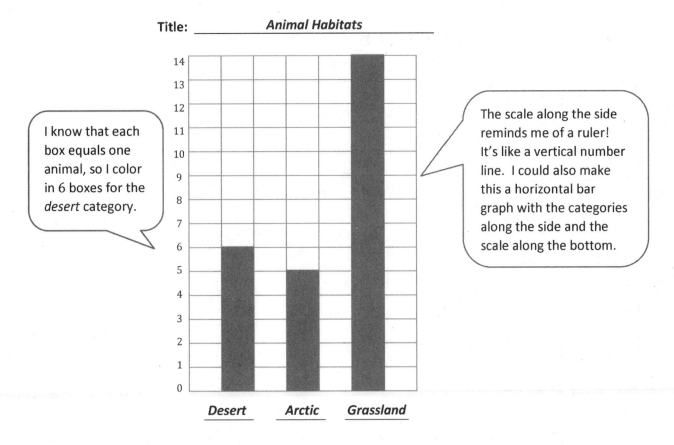

Title: _____ *Animal Habitats* _____

I know that each box equals one animal, so I color in 6 boxes for the *desert* category.

The scale along the side reminds me of a ruler! It's like a vertical number line. I could also make this a horizontal bar graph with the categories along the side and the scale along the bottom.

a.   How many total animals are living in the three habitats?   **25**

$$6 + 5 + 14 = 11 + 14 = 25$$

Lesson 3:   Draw and label a bar graph to represent data; relate the count scale to the number line.

101

© 2018 Great Minds®. eureka-math.org

b.   How many more animals live in the grassland than in the desert and arctic combined?   __3__

$$6 + 5 = 11 \qquad\qquad 14 - 11 = 3$$

> When I combine the number of boxes I colored for *desert* and *arctic*, I count 11. I look at the graph and see that 11 is 3 fewer boxes than 14, which is the number of animals living in the grassland.

c.   If 2 animals were removed from each category, how many animals would there be?   __19__

$$4 + 3 + 12 = 19$$

Lesson 3:   Draw and label a bar graph to represent data; relate the count scale to
the number line.

© 2018 Great Minds®. eureka-math.org

Name _____ Date _____

1. Complete the bar graph below using data provided in the table. Then, answer the questions about the data.

Various Animal Coverings at Jake's Pet Shop			
Fur	Feathers	Shells	Scales
12	9	8	11

Title: _____

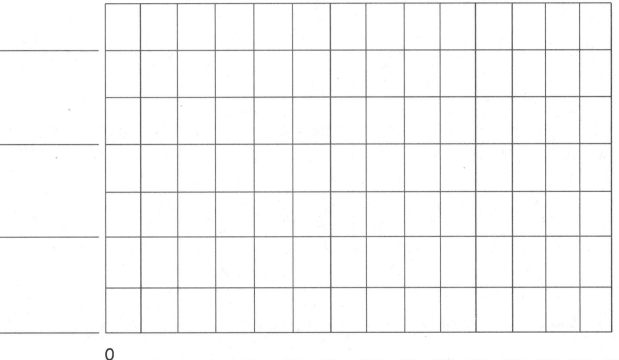

a. How many more animals have fur than shells? ___

b. Which pair of categories has more, fur and feathers or shells and scales? (Circle one.) How much more? ___

c. Write and answer your own comparison question based on the data.

Question: _____

Answer: _____

Lesson 3: Draw and label a bar graph to represent data; relate the count scale to the number line.

103

© 2018 Great Minds®. eureka-math.org

2. Complete the bar graph below using data provided in the table.

City Shelter Animal Diets		
Meat Only	Plants Only	Meat and Plants
ⅢⅢ III	ⅢⅢ IIII	ⅢⅢ ⅢⅢ IIII

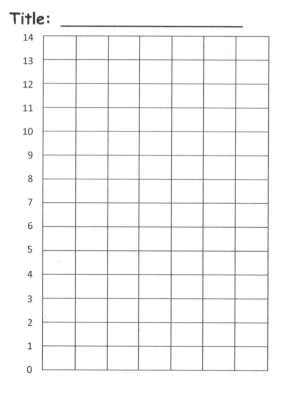

Title: _____

a. How many total animals are in the city shelter? _____

b. How many more meat- and plant-eating animals are there than meat only? _____

c. If 3 animals were removed from each category, how many animals would there be? _____

d. Write your own comparison question based on the data, and answer it.

Question: _____

Answer: _____

Lesson 3:   Draw and label a bar graph to represent data; relate the count scale to the number line.

EUREKA
MATH

Complete the bar graph using the table with the types of bugs Alicia counted in the park. Then, answer the following questions.

Types of Bugs			
Butterflies	Spiders	Bees	Grasshoppers
5	14	12	7

> Before I can record the data, I need to write a title for the graph, label the four categories, and write a number scale at the bottom.

Title: _____ *Types of Bugs* _____

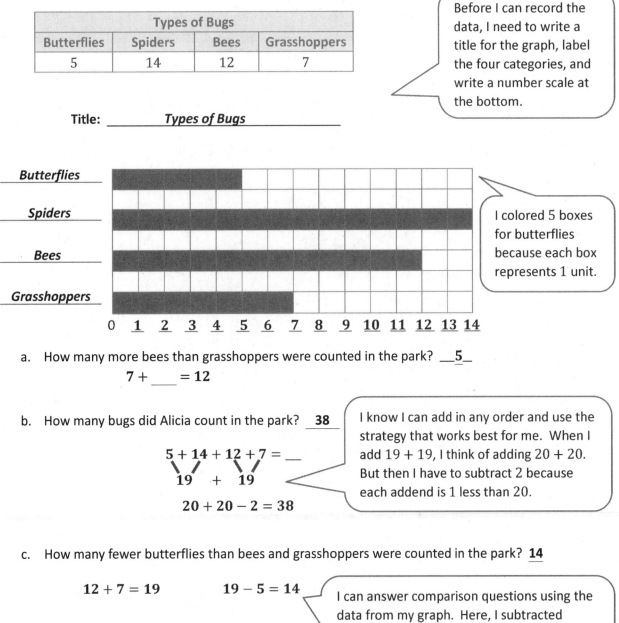

> I colored 5 boxes for butterflies because each box represents 1 unit.

a. How many more bees than grasshoppers were counted in the park? __5__

$$7 + \underline{} = 12$$

b. How many bugs did Alicia count in the park? __38__

$$5 + 14 + 12 + 7 = \underline{}$$
$$19 + 19$$
$$20 + 20 - 2 = 38$$

> I know I can add in any order and use the strategy that works best for me. When I add 19 + 19, I think of adding 20 + 20. But then I have to subtract 2 because each addend is 1 less than 20.

c. How many fewer butterflies than bees and grasshoppers were counted in the park? **14**

$$12 + 7 = 19 \qquad 19 - 5 = 14$$

> I can answer comparison questions using the data from my graph. Here, I subtracted 19 − 5 = 14. In part (a), I thought of the missing part to solve, 7 + _ = 12. I can use both operations!

Name _____   Date _____

1. Complete the bar graph using the table with the types of reptiles at the local zoo.
    Then, answer the following questions.

Types of Reptiles			
Snakes	Lizards	Turtles	Tortoises
13	11	7	8

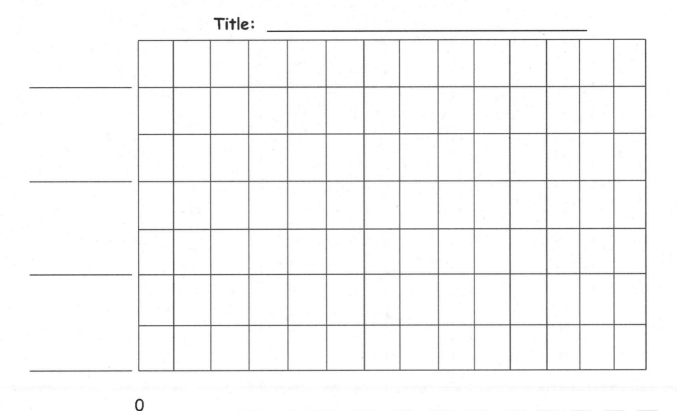

Title: _____

0   __ __ __ __ __ __ __ __ __ __ __ __ __ __ __

a. How many reptiles are at the zoo? _____

b. How many more snakes and lizards than turtles are at the zoo? _____

c. How many fewer turtles and tortoises than snakes and lizards are at the zoo?
    _____

d. Write a comparison question that can be answered using the data on the
    bar graph.
    _____

2. Complete the bar graph with labels and numbers using the number of underwater animals Emily saw while scuba diving.

Underwater Animals			
Sharks	Stingrays	Starfish	Seahorses
6	9	14	13

Title: _____

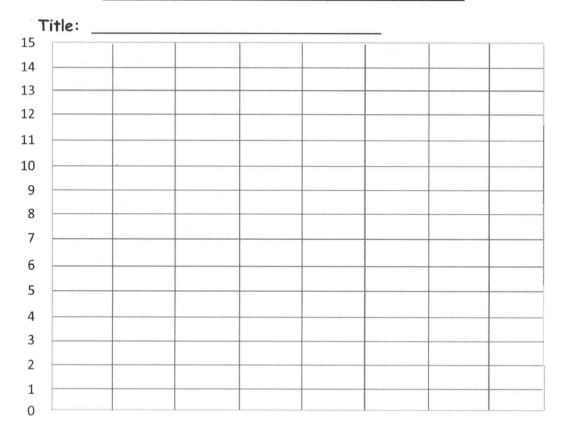

_____ _____ _____ _____

a. How many more starfish than sharks did Emily see? _____

b. How many fewer stingrays than seahorses did Emily see? _____

c. Write a comparison question that can be answered using the data on the bar graph.

_____

Lesson 4:   Draw a bar graph to represent a given data set.

Use the table to complete the bar graph. Then, answer the following questions.

Number of Dimes Donated			
Madison	Ross	Bella	Miguel
15	9	12	11

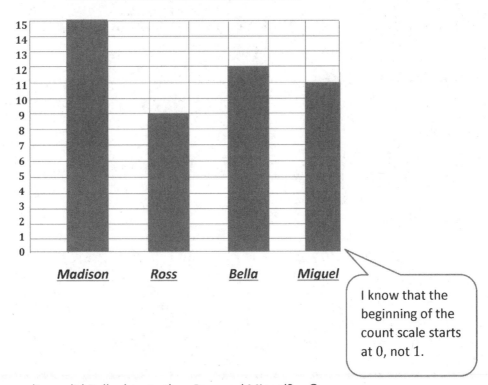

Title: ___*Number of Dimes Donated*___

I know that the beginning of the count scale starts at 0, not 1.

a. How many fewer dimes did Bella donate than Ross and Miguel? __8__

$9 + 11 = 20$ $12 + \_\_\_\_ = 20$

b. How many more dimes are needed for Madison to donate the same as Ross and Bella? __6__

$9 + 12 = 21$ $15 + \_\_\_\_ = 21$

Lesson 5:   Solve word problems using data presented in a bar graph.

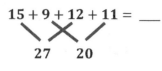

c.   How many dimes were donated in total? __47__

$15 + 9 + 12 + 11 =$ ___

27   20

$27 + 20 = 47$

> I can use mental math to find the total. I can make a ten: $9 + 11 = 20$. It's easy to add the tens and ones when I combine 15 and 12. Then, $27 + 20 = 47$.

d.   Circle the pair that has more dimes, Madison and Ross or Bella and Miguel. How many more?

__1__

$15 + 9 = 24$ $12 + 11 = 23$ $24 - 23 = 1$

Name _____   Date _____

1. Use the table to complete the bar graph. Then, answer the following questions.

Number of Nickels

Justin	Melissa	Meghan	Douglas
13	9	12	7

Title: _____

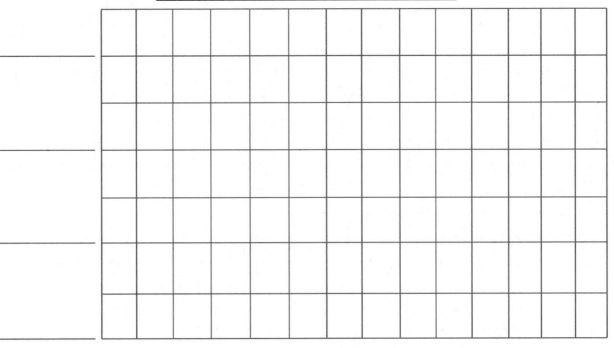

a. How many more nickels does Meghan have than Melissa? _____

b. How many fewer nickels does Douglas have than Justin? _____

c. Circle the pair that has more nickels, Justin and Melissa or Douglas and Meghan.
    How many more? _____

d. What is the total number of nickels if all the students combine all their money?

    _____

2. Use the table to complete the bar graph. Then, answer the following questions.

Dimes Donated

Kylie	Tom	John	Shannon
12	10	15	13

Title: _____

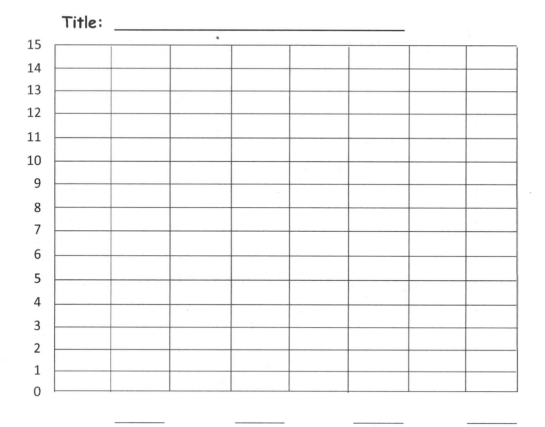

a. How many dimes did Shannon donate? _____

b. How many fewer dimes did Kylie donate than John and Shannon? _____

c. How many more dimes are needed for Tom to donate the same as Shannon and Kylie? _____

d. How many dimes were donated in total? _____

Lesson 5: Solve word problems using data presented in a bar graph.

EUREKA MATH

Count or add to find the total value of each group of coins.

Write the value using the ¢ or $ symbol.

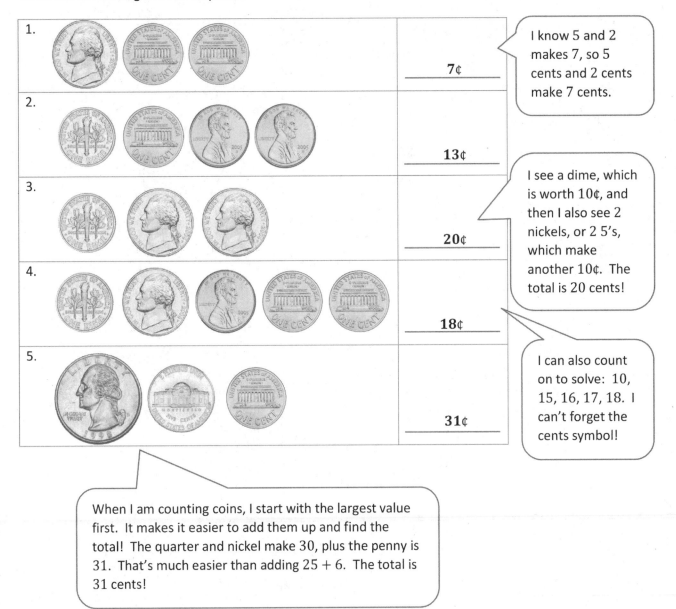

1.	7¢
2.	13¢
3.	20¢
4.	18¢
5.	31¢

I know 5 and 2 makes 7, so 5 cents and 2 cents make 7 cents.

I see a dime, which is worth 10¢, and then I also see 2 nickels, or 2 5's, which make another 10¢. The total is 20 cents!

I can also count on to solve: 10, 15, 16, 17, 18. I can't forget the cents symbol!

When I am counting coins, I start with the largest value first. It makes it easier to add them up and find the total! The quarter and nickel make 30, plus the penny is 31. That's much easier than adding 25 + 6. The total is 31 cents!

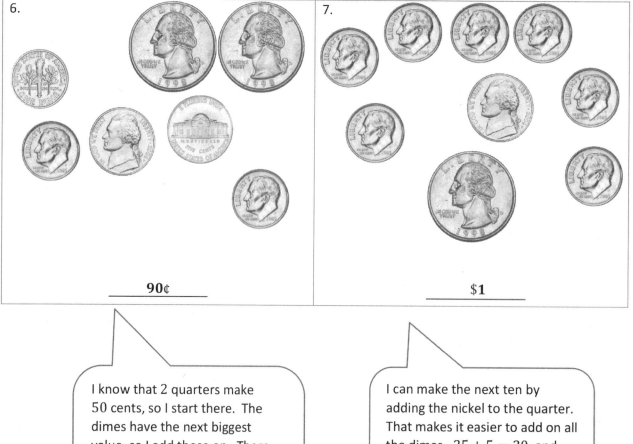

6. **90¢**

7. **$1**

I know that 2 quarters make 50 cents, so I start there. The dimes have the next biggest value, so I add those on. There are 3 dimes, so I add on 30 cents. Then there are 2 nickels, so I add on 10 more cents. The total is 90 cents!

I can make the next ten by adding the nickel to the quarter. That makes it easier to add on all the dimes. $25 + 5 = 30$, and then I skip-count 40, 50, ..., 100. 100 cents is one dollar!

Lesson 6:   Recognize the value of coins and count up to find their total value.

Name _____   Date _____

Count or add to find the total value of each group of coins.

Write the value using the ¢ or $ symbol.

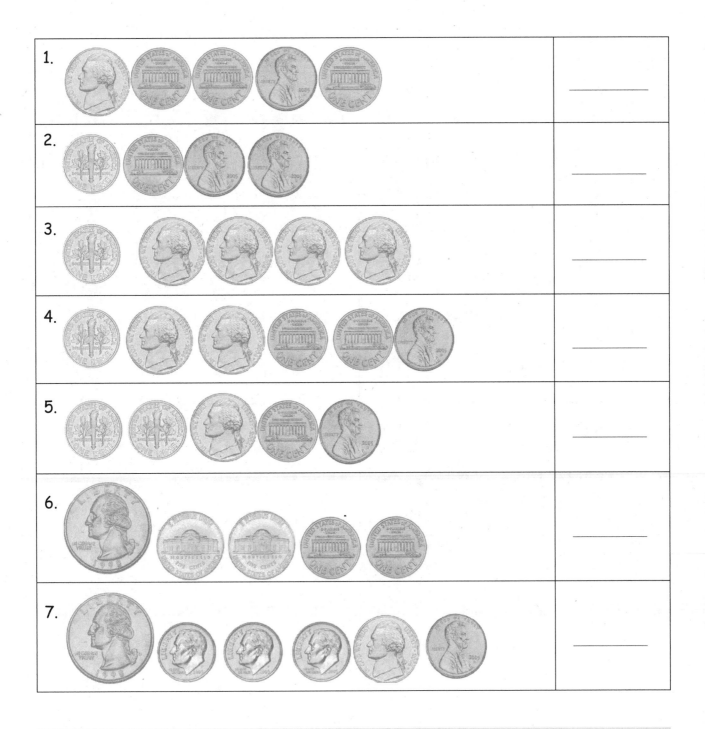

EUREKA MATH

Lesson 6:   Recognize the value of coins and count up to find their total value.

115

© 2018 Great Minds®. eureka-math.org

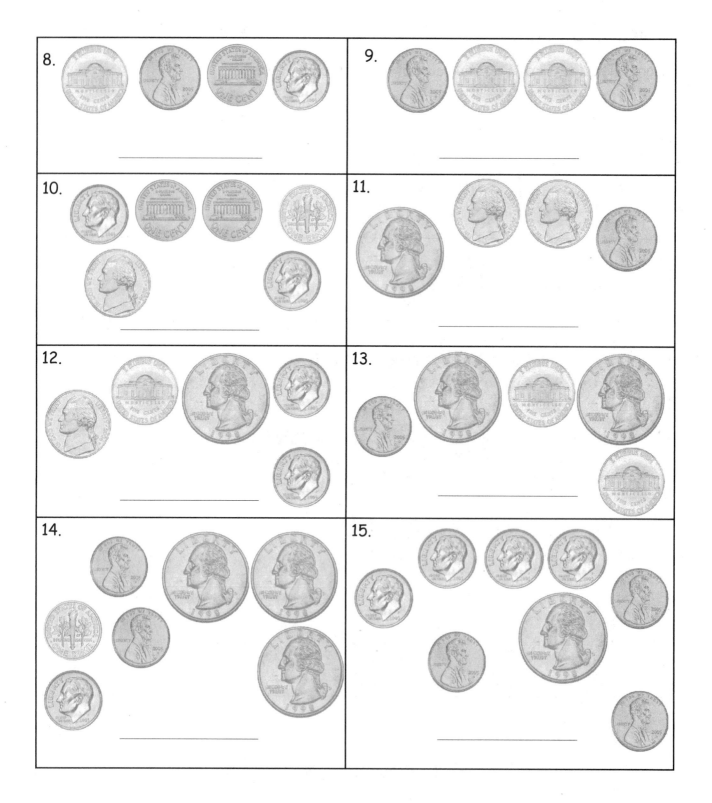

8. _____

9. _____

10. _____

11. _____

12. _____

13. _____

14. _____

15. _____

Lesson 6: Recognize the value of coins and count up to find their total value.

EUREKA MATH

Solve.

Enrique had 2 quarters, 2 dimes, 5 pennies, and 3 nickels in his wallet. Then, he bought a lemonade for 25 cents. How much money did he have left?

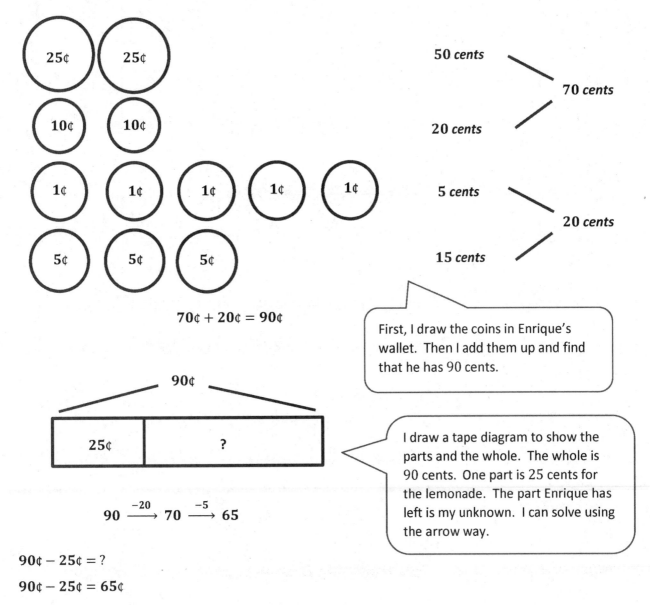

$$70¢ + 20¢ = 90¢$$

First, I draw the coins in Enrique's wallet. Then I add them up and find that he has 90 cents.

I draw a tape diagram to show the parts and the whole. The whole is 90 cents. One part is 25 cents for the lemonade. The part Enrique has left is my unknown. I can solve using the arrow way.

$$90 \xrightarrow{-20} 70 \xrightarrow{-5} 65$$

$90¢ - 25¢ = ?$

$90¢ - 25¢ = 65¢$

Enrique had 65 cents left.

Lesson 7:   Solve word problems involving the total value of a group of coins.

117

EUREKA MATH

© 2018 Great Minds®. eureka-math.org

Name _____   Date _____

Solve.

1. Owen has 4 dimes, 3 nickels, and 16 pennies. How much money does he have?

2. Eli found 1 quarter, 1 dime, and 2 pennies in his desk and 16 pennies and 2 dimes in his backpack. How much money does he have in all?

3. Carrie had 2 dimes, 1 quarter, and 11 pennies in her pocket. Then, she bought a soft pretzel for 35 cents. How much money does Carrie have left?

4. Ethan had 67 cents. He gave 1 quarter and 6 pennies to his sister. How much money does Ethan have left?

5. There are 4 dimes and 3 nickels in Susan's piggy bank. Nevaeh has 17 pennies and 3 nickels in her piggy bank. What is the total value of the money in both piggy banks?

6. Tison had 1 quarter, 4 dimes, 4 nickels, and 5 pennies. He gave 57 cents to his cousin. How much money does Tison have left?

Solve.

Claire has $89. She has 3 more five-dollar bills, 4 more one-dollar bills, and 1 more ten-dollar bill than Trey. How much money does Trey have?

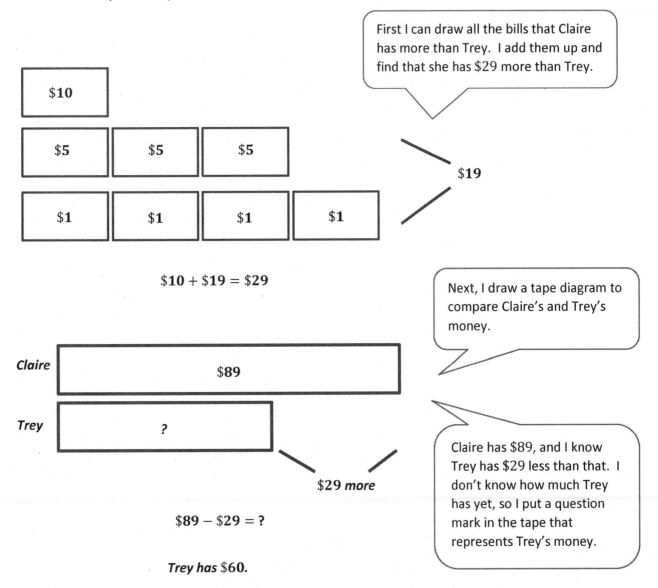

First I can draw all the bills that Claire has more than Trey. I add them up and find that she has $29 more than Trey.

$10 + $19 = $29

Next, I draw a tape diagram to compare Claire's and Trey's money.

Claire has $89, and I know Trey has $29 less than that. I don't know how much Trey has yet, so I put a question mark in the tape that represents Trey's money.

$89 − $29 = ?

Trey has $60.

Lesson 8:   Solve word problems involving the total value of a group of bills.

121

© 2018 Great Minds®. eureka-math.org

Name _____   Date _____

Solve.

1. Mr. Chang has 4 ten-dollar bills, 3 five-dollar bills, and 6 one-dollar bills. How much money does he have in all?

2. At her yard sale, Danielle got 1 twenty-dollar bill and 5 one-dollar bills last week. This week, she got 3 ten-dollar bills and 3 five-dollar bills. What is the total amount she got for both weeks?

3. Patrick has 2 fewer ten-dollar bills than Brenna. Patrick has $64. How much money does Brenna have?

4. On Saturday, Mary Jo received 5 ten-dollar bills, 4 five-dollar bills, and 17 one-dollar bills. On Sunday, she received 4 ten-dollar bills, 5 five-dollar bills, and 15 one-dollar bills. How much more money did Mary Jo receive on Saturday than on Sunday?

5. Alexis has $95. She has 2 more five-dollar bills, 5 more one-dollar bills, and 2 more ten-dollar bills than Kasai. How much money does Kasai have?

6. Kate had 2 ten-dollar bills, 6 five-dollar bills, and 21 one-dollar bills before she spent $45 on a new outfit. How much money was not spent?

Lesson 8:   Solve word problems involving the total value of a group of bills.

© 2018 Great Minds®. eureka-math.org

1. Write another way to make the same total value.

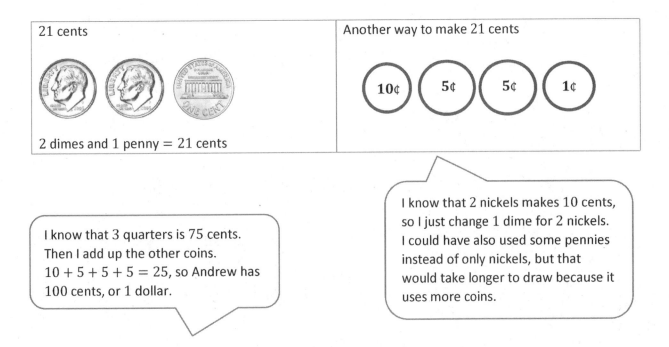

21 cents	Another way to make 21 cents
2 dimes and 1 penny = 21 cents	10¢ 5¢ 5¢ 1¢

I know that 2 nickels makes 10 cents, so I just change 1 dime for 2 nickels. I could have also used some pennies instead of only nickels, but that would take longer to draw because it uses more coins.

I know that 3 quarters is 75 cents. Then I add up the other coins. $10 + 5 + 5 + 5 = 25$, so Andrew has 100 cents, or 1 dollar.

2. Andrew has 3 quarters, 1 dime, 2 nickels, and 5 pennies in his pocket. Write two other coin combinations that would equal the same amount of change.

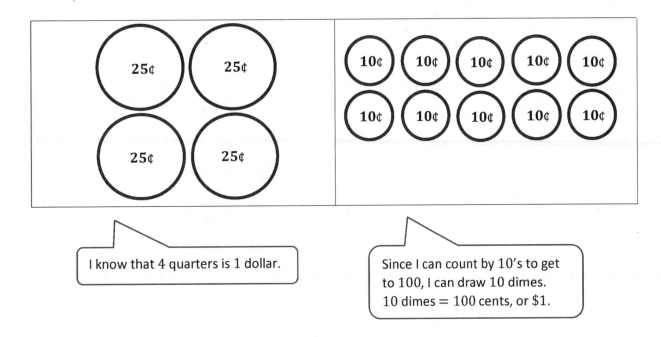

I know that 4 quarters is 1 dollar.

Since I can count by 10's to get to 100, I can draw 10 dimes. 10 dimes = 100 cents, or $1.

Lesson 9: Solve word problems involving different combinations of coins with the same total value.

125

© 2018 Great Minds®. eureka-math.org

Name _____   Date _____

Draw coins to show another way to make the same total value.

1. 25 cents 1 dime 3 nickels is 25 cents.	Another way to make 25 cents:
2. 40 cents 4 dimes make 40 cents.	Another way to make 40 cents:
3. 60 cents 2 quarters and 1 dime makes 60 cents.	Another way to make 60 cents:
4. 80 cents The total value of 3 quarters 1 nickel is 80 cents.	Another way to make 80 cents:

Lesson 9:   Solve word problems involving different combinations of coins with the same
total value.

127

© 2018 Great Minds®. eureka-math.org

5. Samantha has 67 cents in her pocket. Write two coin combinations she could have that would equal the same amount.

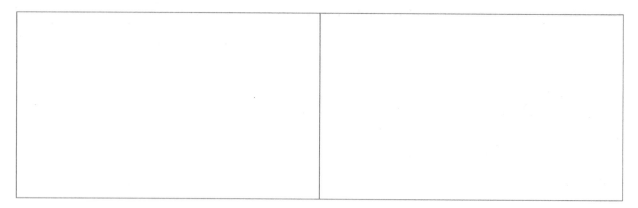

6. The store clerk gave Jeremy 2 quarters, 3 nickels, and 4 pennies. Write two other coin combinations that would equal the same amount of change.

7. Chelsea has 10 dimes. Write two other coin combinations she could have that would equal the same amount.

Lesson 9: Solve word problems involving different combinations of coins with the same total value.

© 2018 Great Minds®. eureka-math.org

EUREKA MATH

1. Ana showed 30 cents two ways. Circle the way that uses the fewest coins.

a. [nickel nickel dime dime]   b. [dime dime dime] (circled)

What two coins from part (a) were changed for one coin in part (b)?

Ana changed 2 nickels for 1 dime.

> Ana had 2 nickels, which equal 10 cents, so she was able to change them for 1 dime.

2. Show 74 cents two ways. Use the fewest possible coins on the right below.

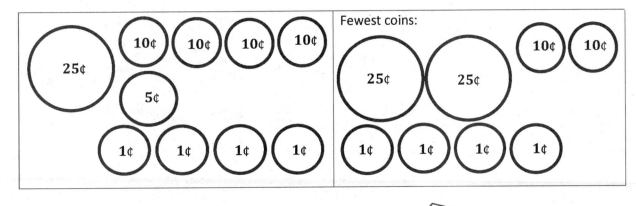

Fewest coins:

> For the fewest coins, I start with the quarter because it has the highest value. 25, 50, 75. Oops, 3 quarters is too much! I'll stop at 50 cents. Now, I add on the next highest value, dimes. 60, 70. I need 4 cents more, so I add 4 pennies.

3. Shelby made a mistake when asked for two ways to show 66¢. Circle her mistake, and explain what she did wrong.

	Fewest coins:
2 quarters, 1 dime, 1 nickel, 1 penny	6 dimes, 1 nickel, 1 penny

The first combination is the fewest coins. Since 2 quarters have the same value as 5 dimes, Shelby only

needs 5 coins to make 66¢. Her second combination uses 8 coins.

Name _____ Date _____

1. Tara showed 30 cents two ways. Circle the way that uses the fewest coins.

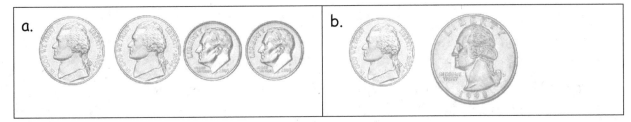

What coins from (a) were changed for one coin in (b)?

_____

2. Show 40¢ two ways. Use the fewest possible coins on the right below.

	Fewest coins:

3. Show 55¢ two ways. Use the fewest possible coins on the right below.

	Fewest coins:

4. Show 66¢ two ways. Use the fewest possible coins on the right below.

	Fewest coins:

5. Show 80¢ two ways. Use the fewest possible coins on the right below.

	Fewest coins:

6. Show $1 two ways. Use the fewest possible coins on the right below.

	Fewest coins:

7. Tara made a mistake when asked for two ways to show 91¢. Circle her mistake, and explain what she did wrong.

	Fewest coins:
3 quarters, 1 dime, 1 nickel, and 1 penny	9 dimes and 1 penny

_____

_____

_____

Lesson 10:   Use the fewest number of coins to make a given value.

1. Count up using the arrow way to complete each number sentence. Then, use coins to check your answers, if possible.

$$65¢ + \underline{\textbf{35¢}} = 100¢$$

$$65 \xrightarrow{+5} 70 \xrightarrow{+30} 100$$

> I start at 65 cents and add 5 more to get to the next 10, which is 70 cents. I know I need 30 more cents to get to 100 cents, or \$1. $5 + 30 = 35$, so the missing part is 35 cents.

2. Solve using the arrow way and a number bond.

$$22¢ + \underline{\textbf{78¢}} = 100¢$$

$$22 \xrightarrow{+8} 30 \xrightarrow{+70} 100$$

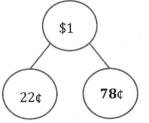

> I use the number bond to show that the whole is \$1, and there are two parts. The part I know already is 22 cents. After I solve using the arrow way, I can fill in the missing part, which is 78 cents.

$$100¢ - 65¢ = \underline{\textbf{35¢}}$$

$$100 \xrightarrow{-60} 40 \xrightarrow{-5} 35$$

> I use the arrow way to subtract, too! If I buy something for 65 cents, and I give the cashier 1 dollar, I will get 35 cents in change!

Name _____   Date _____

1. Count up using the arrow way to complete each number sentence. Then, use coins to check your answers, if possible.

    a. 25¢ + _____ = 100¢   b. 45¢ + _____ = 100¢

    $25 \xrightarrow{+5} \underline{\quad} \xrightarrow{+\underline{\quad}} 100$

    c. 62¢ + _____ = 100¢   d. _____ + 79¢ = 100¢

2. Solve using the arrow way and a number bond.

    a. 19¢ + _____ = 100¢

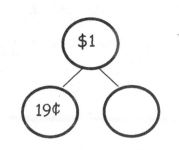

    b. 77¢ + _____ = 100¢

    c. 100¢ – 53¢ = _____

3. Solve.

   a. _____ + 38¢ = 100¢

   b. 100¢ – 65¢ = _____

   c. 100¢ – 41¢ = _____

   d. 100¢ – 27¢ = _____

   e. 100¢ – 14¢ = _____

Lesson 11: Use different strategies to make $1 or make change from $1.

© 2018 Great Minds®. eureka-math.org

Maria has 1 quarter, 8 pennies, 4 nickels, and 1 dime. She needs $1 to ride the bus. How much should Maria borrow from her mom?

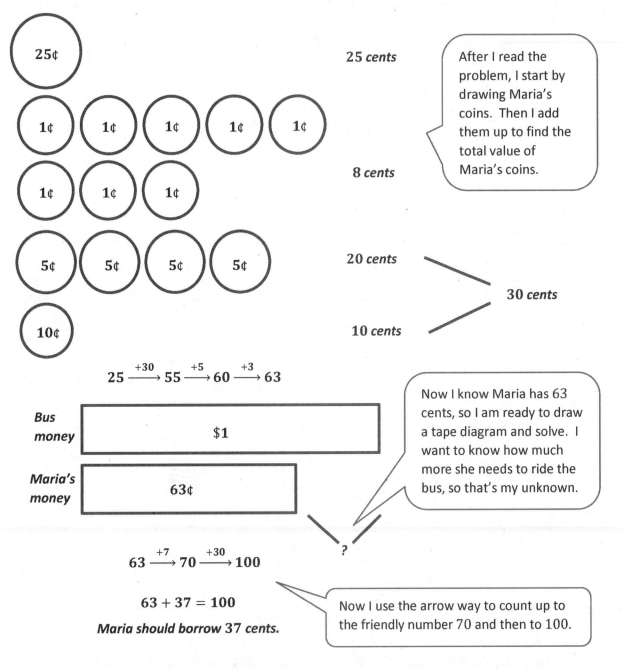

25 *cents*

After I read the problem, I start by drawing Maria's coins. Then I add them up to find the total value of Maria's coins.

8 *cents*

20 *cents*

30 *cents*

10 *cents*

$$25 \xrightarrow{+30} 55 \xrightarrow{+5} 60 \xrightarrow{+3} 63$$

Bus money $1

Now I know Maria has 63 cents, so I am ready to draw a tape diagram and solve. I want to know how much more she needs to ride the bus, so that's my unknown.

Maria's money   63¢

?

$$63 \xrightarrow{+7} 70 \xrightarrow{+30} 100$$

$$63 + 37 = 100$$

Maria should borrow 37 cents.

Now I use the arrow way to count up to the friendly number 70 and then to 100.

EUREKA MATH®

© 2018 Great Minds®. eureka-math.org

Name _____   Date _____

Solve using the arrow way, a number bond, or a tape diagram.

1. Kevin had 100 cents. He spent 3 dimes, 3 nickels, and 4 pennies on a balloon.
   How much money does he have left?

2. Colin bought a postcard for 45 cents. He gave the cashier $1. How much change did
   he receive?

3. Eileen spent 75 cents of her dollar at the market. How much money does she
   have left?

4. The puzzle Casey wants costs $1. She has 6 nickels, 1 dime, and 11 pennies.
   How much more money does she need to buy the puzzle?

5. Garret found 19 cents in the sofa and 34 cents under his bed. How much more
   money will he need to find to have $1?

6. Kelly has 38 fewer cents than Molly. Molly has $1. How much money does
   Kelly have?

7. Mario has 41 more cents than Ryan. Mario has $1. How much money does
   Ryan have?

Lesson 12:   Solve word problems involving different ways to make change from $1.

EUREKA
MATH

James had 1 quarter, 1 dime, and 12 pennies. He found 3 coins under his bed. Now he has 77 cents. What 3 coins did he find?

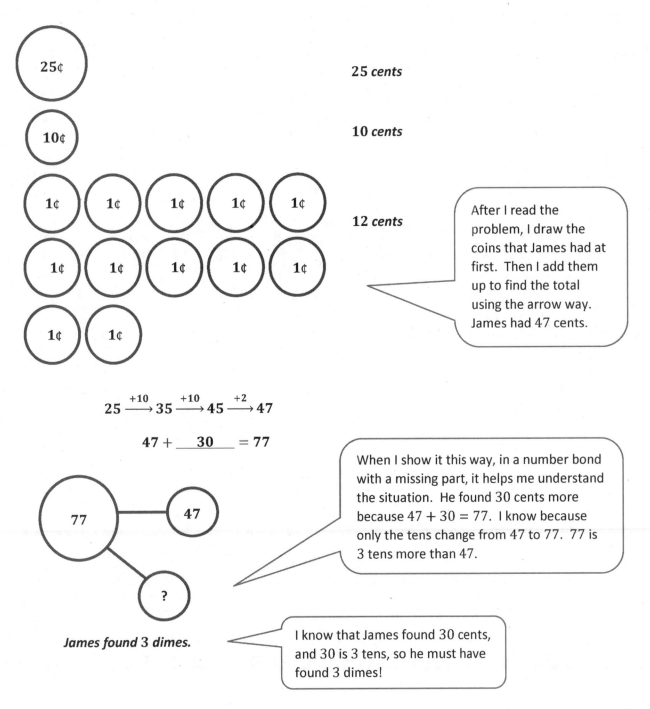

25 *cents*

10 *cents*

12 *cents*

> After I read the problem, I draw the coins that James had at first. Then I add them up to find the total using the arrow way. James had 47 cents.

$$25 \xrightarrow{+10} 35 \xrightarrow{+10} 45 \xrightarrow{+2} 47$$

$$47 + \underline{\quad 30 \quad} = 77$$

> When I show it this way, in a number bond with a missing part, it helps me understand the situation. He found 30 cents more because $47 + 30 = 77$. I know because only the tens change from 47 to 77. 77 is 3 tens more than 47.

James found 3 dimes.

> I know that James found 30 cents, and 30 is 3 tens, so he must have found 3 dimes!

Lesson 13: Solve two-step word problems involving dollars or cents with totals within $100 or $1.

141

© 2018 Great Minds®. eureka-math.org

Name _____   Date _____

1. Kelly bought a pencil sharpener for 47 cents and a pencil for 35 cents. What was her change from $1?

2. Hae Jung bought a pretzel for 3 dimes and a nickel. She also bought a juice box. She spent 92 cents. How much was the juice box?

3. Nolan has 1 quarter, 1 nickel, and 21 pennies. His brother gave him 2 coins. Now he has 86 cents. What 2 coins did his brother give him?

EUREKA
MATH

Lesson 13:   Solve two-step word problems involving dollars or cents with totals within
               $100 or $1.

143

© 2018 Great Minds®. eureka-math.org

4. Monique saved 2 ten-dollar bills, 4 five-dollar bills, and 15 one-dollar bills. Harry saved $16 more than Monique. How much money does Harry have saved?

5. Ryan went shopping with 3 twenty-dollar bills, 3 ten-dollar bills, 1 five-dollar bill, and 9 one-dollar bills. He spent 59 dollars on a video game. How much money does he have left?

6. Heather had 3 ten-dollar bills and 4 five-dollar bills left after buying a new pair of sneakers for $29. How much money did she have before buying the sneakers?

144   Lesson 13:   Solve two-step word problems involving dollars or cents with totals within $100 or $1.

© 2018 Great Minds®. eureka-math.org

EUREKA MATH

1. Measure these objects found in your home with an inch tile. Record the measurements in the table provided.

Object	Measurement
Length of a hairbrush	4 inches
Height of a milk carton	10 inches
Length of the oven	27 inches

I put the tile at one end of the milk carton and make a mark where the tile begins and ends. Then, I move the tile forward and place the edge right on top of the previous hash mark.

Since I can't draw on an oven, I used the tip of my pencil to remind me where to place my inch tile each time. The spaces between my hash marks are the same length each time.

I leave no spaces between my inch tile and the hash marks I draw!

I use the mark and move forward strategy when I am measuring my little hairbrush with my red inch tile. I put my inch tile down touching the endpoint of the hairbrush. Then I make a mark where the inch tile ends so I know where to place it when I move it over.

I count the spaces between my hash marks to see how many inches long my hairbrush is. My hairbrush is almost 4 inches, so I can say it's about 4 inches.

© 2018 Great Minds®. eureka-math.org

2. Charlene measures her pencil with her inch tile. She marks off where each inch ends so she knows where to place the tile. Charlene says the pencil is 4 inches long.

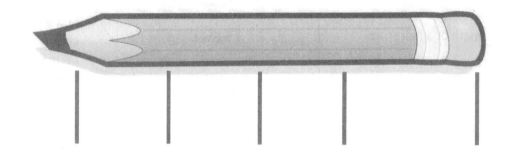

Is Charlene's measurement correct? Explain your answer.

It looks like Charlene did not start her measurement in the correct place. The first hash mark is not

lined up with the endpoint of the pencil. It also looks like she was not careful with her measuring

because the last hash mark looks farther than an inch from the one before. She is not correct.

3. Use your inch tile to measure the pencil. How many inch tiles long is the pencil? Explain how you know.

I was very careful to start at the tip of the pencil. I made a hash mark at the endpoint of the

pencil. I used the mark and move forward strategy and was careful not to leave any space

between my tile and my hash marks. The pencil is about 5 inches long.

Lesson 14:   Connect measurement with physical units by using iteration with an
                                inch tile to measure.

Name _____ Date _____

1. Measure these objects found in your home with an inch tile. Record the measurements in the table provided.

Object	Measurement
Length of a kitchen fork	
Height of a juice glass	
Length across the center of a plate	
Length of the refrigerator	
Length of a kitchen drawer	
Height of a can	
Length of a picture frame	
Length of a remote control	

Lesson 14:   Connect measurement with physical units by using iteration with an inch tile to measure.

147

© 2018 Great Minds®. eureka-math.org

2. Norberto begins measuring his pen with his inch tile. He marks off where each tile ends. After two times, he decides this process is taking too long and starts to guess where the tile would end and then marks it.

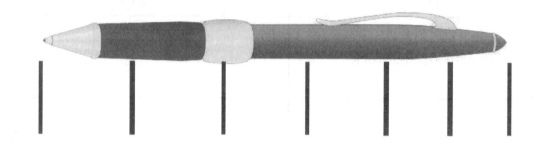

Explain why Norberto's answer will not be correct.

_____

_____

_____

3. Use your inch tile to measure the pen. How many inch tiles long is the pen?

Lesson 14:   Connect measurement with physical units by using iteration with an inch tile to measure.

© 2018 Great Minds®. eureka-math.org

1. Measure the length of the object with your ruler, and then use your ruler to draw a line equal to the length of the object in the space provided.

    a.   A toothbrush is ____6____ inches.

    When I measure my toothbrush, I line up the end of the toothbrush with the 0 on my ruler. The end of the handle is even with the 0 on my ruler.

    b.   Draw a line that is the same length as the toothbrush.

    _____

    When I draw my line, I start at 0 and stop after 6 length units. My line is 6 inches long!

2. Measure another household object.

    a.   A _____**bar of soap**_____ is __4__ inches.

    b.   Draw a line that is the same length as the _____**bar of soap**_____.

    _____

3.

    a.   Which object was longer? _____**toothbrush**_____

    I can tell the toothbrush is longer just by looking at the objects or the lines I drew. But to know how much longer it is, I can subtract! $6 - 4 = 2$, so the soap is 2 inches shorter.

    b.   Which object was shorter? _____**bar of soap**_____

    c.   The difference between the longer object and the shorter object is __2__ inches.

4. Measure and label the length of each side of the shape in inches using your ruler.

   a.   The longest side of the rectangle is __**4**__ inches.

> To find the difference, I just subtract! $4 - 1 = 3$

   b.   The shortest side of the rectangle is __**1**__ inch.

   c.   The longest side of the rectangle is __**3**__ inches longer than the shortest side of the rectangle.

> Measuring objects with my ruler is so much quicker than using an inch tile! It's like all the tiles are connected!

    Lesson 15:   Apply concepts to create inch rulers; measure lengths using inch rulers.

Name _____   Date _____

Measure the length of each household object with your ruler, and then use your ruler to draw a line equal to the length of each object in the space provided.

1. a. A dinner fork is _____ inches.
    b. Draw a line that is the same length as the fork.

2. a. A tablespoon is _____ inches.
    b. Draw a line that is the same length as the tablespoon.

Measure two other household objects.

3. a. _____ is _____ inches.
    b. Draw a line that is the same length as the _____.

4. a. _____ is _____ inches.
    b. Draw a line that is the same length as the _____.

5. a. What was the longest object you measured? _____
    b. What was the shortest object you measured? _____
    c. The difference between the longest object and the shortest object is _____ inches.

Lesson 15:   Apply concepts to create inch rulers; measure lengths using inch rulers.

151

© 2018 Great Minds®. eureka-math.org

6. Measure and label the length of each side of each shape in inches using your ruler.

a. The longer side of the rectangle is _____ inches.

b. The shorter side of the rectangle is _____ inches.

c. The longer side of the rectangle is _____ inches longer than the shorter side of the rectangle.

d. The shortest side of the trapezoid is _____ inches.

e. The longest side of the trapezoid is _____ inches.

f. The longest side of the trapezoid is _____ inches longer than the shortest side.

g. Each side of the hexagon is _____ inches.

h. The total length around the hexagon is _____ inches.

| | 1 | 2 | 3 | 4 | 5 | |

Lesson 15:   Apply concepts to create inch rulers; measure lengths using inch rulers.

1. Circle the unit that would best measure each object.

Length of a window	inch / foot / yard
Height of an office building	inch / foot / yard
Length of a shoe	inch / foot / yard

> I have to think about how long each object is. If it is very, very long, then I know I should use yards to measure because it is more efficient. It would take a very long time to measure an office building in inches, and that means you could make a lot more mistakes!

> I can picture a yardstick in my mind. I know that an airplane is way longer! I think the guitar is about the length of a yardstick because I can hold it in my arms the same way I can hold a yardstick.

2. Circle the correct estimate for each object.

a. The length of an airplane is more than / less than / about the same as the length of a yardstick.

b. The length of a guitar is more than / less than / about the same as the length of a yardstick.

c. The height of a coffee mug is more than / less than / about the same as the length of a 12-inch ruler.

3. Name 3 objects that you find outside. Decide which unit you would use to measure that object. Record it in the chart in a full statement.

Object	Unit
oak tree	I would use _____*yards*_____ to measure the height of an ___*oak tree*___ .
flower	*I would use inches to measure the height of a flower.*
park bench	*I would use feet to measure the height of a park bench.*

I tried to choose objects that I measure in different units. The tree is big so that works for yards. The park bench could also be measured in yards, but if I measure it in feet, I can give a more accurate measurement.

Name _____   Date _____

1. Circle the unit that would best measure each object.

Height of a door	inch / foot / yard
Textbook	inch / foot / yard
Pencil	inch / foot / yard
Length of a car	inch / foot / yard
Length of your street	inch / foot / yard
Paint brush	inch / foot / yard

2. Circle the correct estimate for each object.

a. The height of a flagpole is more than / less than / about the same as the length of a yardstick.

b. The width of a door is more than / less than / about the same as the length of a yardstick.

c. The length of a laptop computer is more than / less than / about the same as the length of a 12-inch ruler.

d. The length of a cell phone is more than / less than / about the same as the length of a 12-inch ruler.

3. Name 3 objects in your classroom. Decide which unit you would use to measure that object. Record it in the chart in a full statement.

Object	Unit
a.	I would use _____ to measure the length of _____.
b.	
c.	

4. Name 3 objects in your home. Decide which unit you would use to measure that object. Record it in the chart in a full statement.

Object	Unit
a.	I would use _____ to measure the length of _____
b.	
c.	

Lesson 16:   Measure various objects using inch rulers and yardsticks.

Estimate the length of each item by using a mental benchmark. Then, measure the item using feet, inches, or yards.

Item	Mental Benchmark	Estimation	Actual Length
Length of a car	*Yardstick or width of a door*	*6 yards*	*5 yards*
Length of the kitchen sink	*Piece of paper*	*2 feet*	*almost 3 feet*
Length of a pen cap	*Quarter*	*1 inch*	*about an inch*

I choose to use the yardstick as my mental benchmark to estimate the length of the car because the car is very long.

I use the paper to estimate the length of the sink because a piece of paper is my mental benchmark for a foot.

I am so close on my estimate of the length of the pen cap! It is easy to picture it next to the quarter, so I estimate 1 inch. The pen cap is just a little longer than 1 inch, so it's about 1 inch.

Name _____ Date _____

Estimate the length of each item by using a mental benchmark. Then, measure the item using feet, inches, or yards.

Item	Mental Benchmark	Estimation	Actual Length
a. Length of a bed			
b. Width of a bed			
c. Height of a table			
d. Length of a table			
e. Length of a book			

Lesson 17: Develop estimation strategies by applying prior knowledge of length and using mental benchmarks.

159

Item	Mental Benchmark	Estimation	Actual Length
f. Length of your pencil			
g. Length of a refrigerator			
h. Height of a refrigerator			
i. Length of a sofa			

Lesson 17: Develop estimation strategies by applying prior knowledge of length and using mental benchmarks.

1. Measure the lines in inches and centimeters. Round the measurements to the nearest inch or centimeter.

_____5_____ centimeters   _____2_____ inches

> Centimeters are smaller, so it takes more of them to cover the length of the line.

2.

    a. Draw a line that is 3 centimeters in length.

    b. Draw a line that is 3 inches in length.

> An inch is longer than a centimeter, so of course my line that is 3 inches is longer than my line that is 3 centimeters!

3. Sam drew a line that is 11 centimeters long. Susan drew a line that is 8 inches long. Susan thinks her line is shorter than Sam's because 8 is less than 11. Explain why Susan's reasoning is incorrect.

Susan's reasoning is incorrect because inches are longer than centimeters. You have to look

at the unit to figure out which line will be longer. An inch is a larger length unit, so Susan's

line is longer even though 8 is a smaller number.

 EUREKA MATH®   Lesson 18:   Measure an object twice using different length units and compare;   161
                                           relate measurement to unit size.

© 2018 Great Minds®. eureka-math.org

Name _____   Date _____

Measure the lines in inches and centimeters. Round the measurements to the nearest inch or centimeter.

1. _____

    _____ cm   _____ in

2. _____

    _____ cm   _____ in

3. _____

    _____ cm   _____ in

4. _____

    _____ cm   _____ in

EUREKA
MATH

Lesson 18:   Measure an object twice using different length units and compare; relate measurement to unit size.

163

© 2018 Great Minds®. eureka-math.org

5. a Draw a line that is 5 centimeters in length.

   b. Draw a line that is 5 inches in length.

6. a. Draw a line that is 7 inches in length.

   b. Draw a line that is 7 centimeters in length.

7. Takeesha drew a line 9 centimeters long. Damani drew a line 4 inches long.
   Takeesha says her line is longer than Damani's because 9 is greater than 4. Explain
   why Takeesha might be wrong.

   _____

   _____

   _____

8. Draw a line that is 9 centimeters long and a line that is 4 inches long to prove that
   Takeesha is wrong.

Lesson 18:   Measure an object twice using different length units and compare;
                       relate measurement to unit size.

1. Measure each set of lines in inches, and write the length on the line. Complete the comparison sentence.

Line A

_____ **2 inches** _____

Line B

_____ **6 inches** _____

Line A measured about __2__ inches.   Line B measured about __6__ inches.

Line B is about __4__ inches longer than Line A.

> To compare the difference in length, I can subtract $6 - 2 = 4$, or I can say $2 + 4 = 6$. Either way, I know that the difference is 4 inches!

2. Solve. Check your answers with a related addition or subtraction sentence.

    a.   9 inches − 7 inches = __2__ inches

    __2__ inches + 7 inches = 9 inches

    > I think of a number bond. Since I know the total and one part, I can figure out the other part. I can think of addition or subtraction to solve!

    b.   9 centimeters + __7__ centimeters = 16 centimeters

    16 centimeters − 7 centimeters = 9 centimeters

Lesson 19: Measure to compare the differences in lengths using inches, feet, and yards.

165

© 2018 Great Minds®. eureka-math.org

Name _____   Date _____

Measure each set of lines in inches, and write the length on the line. Complete the comparison sentence.

1. Line A   _____

    Line B _____

    Line A measured about _____ inches.   Line B measured about _____ inches.

    Line A is about _____ inches **longer** than Line B.

2. Line C   _____

    Line D   _____

    Line C measured about _____ inches.   Line D measured about _____ inches.

    Line D is about _____ inches **shorter** than Line C.

3. Solve. Check your answers with a related addition or subtraction sentence.

    a. 8 inches - 5 inches = _____ inches

        _____ inches + 5 inches = 8 inches

b. 8 centimeters + _____ centimeters = 19 centimeters

c. 17 centimeters - 8 centimeters = _____ centimeters

d. _____ centimeters + 6 centimeters = 18 centimeters

e. 2 inches + _____ inches = 7 inches

f. 12 inches - _____ = 8 inches

Lesson 19:   Measure to compare the differences in lengths using inches, feet, and
                       yards.

© 2018 Great Minds®. eureka-math.org

Solve using tape diagrams. Use a symbol for the unknown.

1. Angela knitted 18 inches of a scarf. She wants her scarf to be 1 yard long. How many more inches does Angela need to knit?

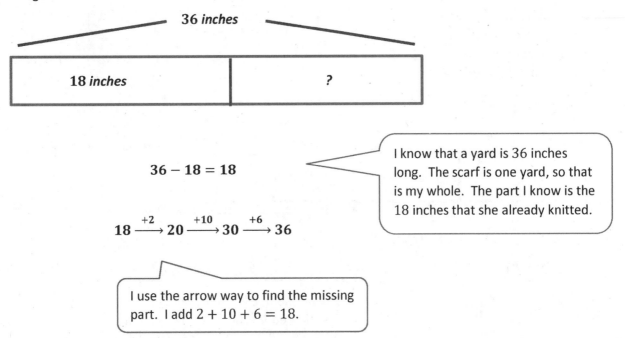

$$36 - 18 = 18$$

$$18 \xrightarrow{+2} 20 \xrightarrow{+10} 30 \xrightarrow{+6} 36$$

I know that a yard is 36 inches long. The scarf is one yard, so that is my whole. The part I know is the 18 inches that she already knitted.

I use the arrow way to find the missing part. I add $2 + 10 + 6 = 18$.

*Angela needs to knit **18** more inches to finish her scarf.*

EUREKA
MATH®

Lesson 20: Solve two-digit addition and subtraction word problems involving length by using tape diagrams and writing equations to represent the problem.

© 2018 Great Minds®. eureka-math.org

169

2. The total length of all three sides of a triangle is 100 feet. Two sides of the triangle are the same length. One of the equal sides measures 40 feet. What is the length of the side that is not equal?

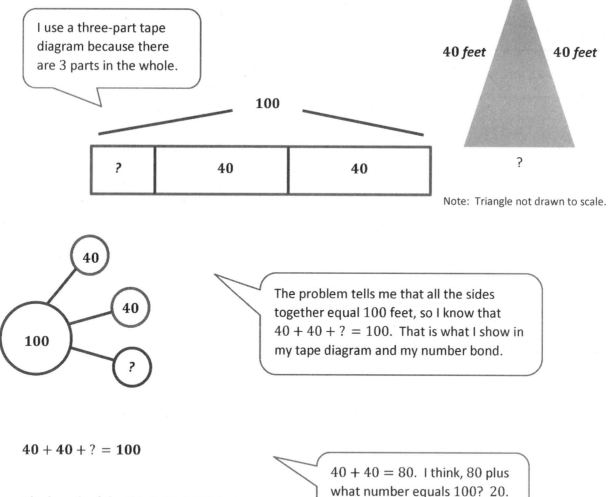

Note: Triangle not drawn to scale.

$40 + 40 + ? = 100$

The length of the third side is 20 feet.

Lesson 20:   Solve two-digit addition and subtraction word problems involving length by using tape diagrams and writing equations to represent the problem.
© 2018 Great Minds®. eureka-math.org

Name _____ Date _____

Solve using tape diagrams. Use a symbol for the unknown.

1. Luann has a piece of ribbon that is 1 yard long. She cuts off 33 inches to tie a gift box. How many inches of ribbon are not used?

2. Elijah runs 68 yards in a 100-yard race. How many more yards does he have to run?

3. Chris has a 57-inch piece of string and another piece that is 15 inches longer than the first. What is the total length of both strings?

EUREKA
MATH

Lesson 20:   Solve two-digit addition and subtraction word problems involving
length by using tape diagrams and writing equations to represent
the problem.
© 2018 Great Minds®. eureka-math.org

171

4. Janine knitted 12 inches of a scarf on Friday and 36 inches on Saturday. She wants the scarf to be 72 inches long. How many more inches does she need to knit?

5. The total length of all three sides of a triangle is 120 feet. Two sides of the triangle are the same length. One of the equal sides measures 50 feet. What is the length of the side that is not equal?

?

6. The length of one side of a square is 3 yards. What is the combined length of all four sides of the square?

Lesson 20:   Solve two-digit addition and subtraction word problems involving
              length by using tape diagrams and writing equations to represent
              the problem.
© 2018 Great Minds®. eureka-math.org

Find the value of the point on each part of the meter strip marked by a letter. For each number line, one unit is the distance from one hash mark to the next. (Note: Number lines not drawn to scale.)

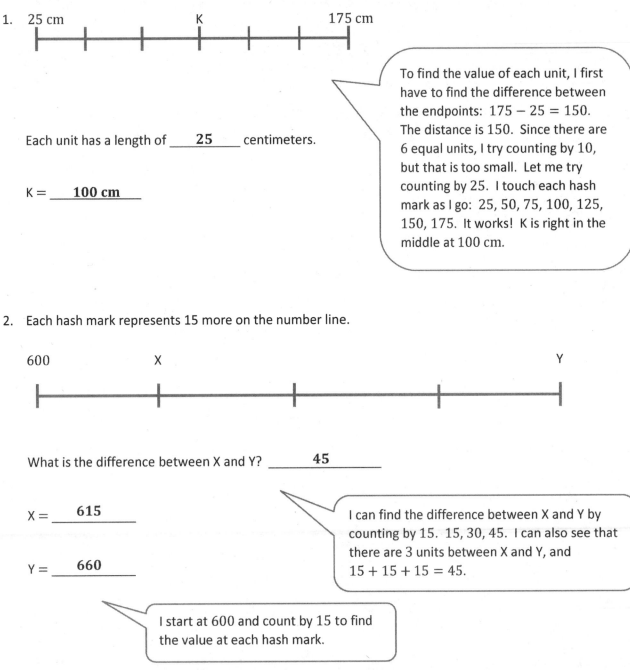

1. 25 cm   K   175 cm

Each unit has a length of _____**25**_____ centimeters.

K = _____**100 cm**_____

> To find the value of each unit, I first have to find the difference between the endpoints: $175 - 25 = 150$. The distance is 150. Since there are 6 equal units, I try counting by 10, but that is too small. Let me try counting by 25. I touch each hash mark as I go: $25, 50, 75, 100, 125, 150, 175$. It works! K is right in the middle at 100 cm.

2. Each hash mark represents 15 more on the number line.

600   X   Y

What is the difference between X and Y? _____**45**_____

X = _____**615**_____

Y = _____**660**_____

> I can find the difference between X and Y by counting by 15. $15, 30, 45$. I can also see that there are 3 units between X and Y, and $15 + 15 + 15 = 45$.

> I start at 600 and count by 15 to find the value at each hash mark.

     Lesson 21:   Identify unknown numbers on a number line diagram by using the distance between numbers and reference points.   **173**

© 2018 Great Minds®. eureka-math.org

Name _____   Date _____

Find the value of the point on each part of the meter strip marked by a letter.
For each number line, one unit is the distance from one hash mark to the next.

1.

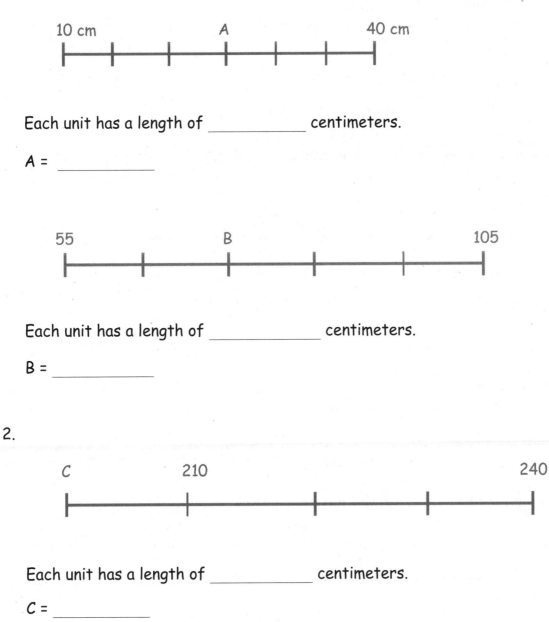

Each unit has a length of _____ centimeters.

A = _____

Each unit has a length of _____ centimeters.

B = _____

2.

Each unit has a length of _____ centimeters.

C = _____

EUREKA
MATH

Lesson 21:   Identify unknown numbers on a number line diagram by using the
               distance between numbers and reference points.   175

© 2018 Great Minds®. eureka-math.org

3. Each hash mark represents 5 more on the number line.

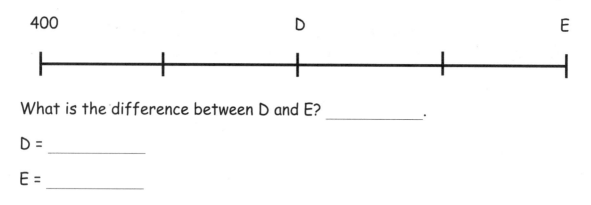

400   D   E

What is the difference between D and E? _____.

D = _____

E = _____

4. Each hash mark represents 10 more on the number line.

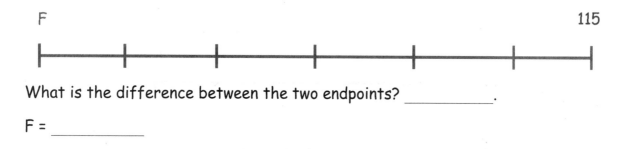

F   115

What is the difference between the two endpoints? _____.

F = _____

5. Each hash mark represents 10 more on the number line.

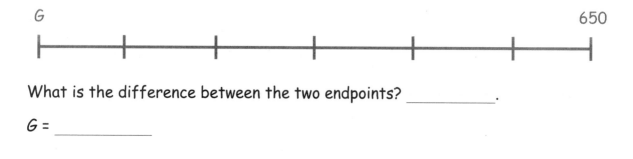

G   650

What is the difference between the two endpoints? _____.

G = _____

Lesson 21:   Identify unknown numbers on a number line diagram by using the
              distance between numbers and reference points.

1. Each unit length on both number lines is 20 feet. (Note: The number lines are not drawn to scale.)

    a. Show 60 feet more than 80 feet on the number line.

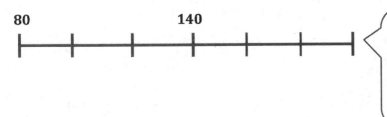

I can show 60 more feet on the number line by labeling the endpoint on the left 80 and then counting on 20, 40, 60. It is the same as adding $80 + 60$.

    b. Write an addition sentence to match the number line.

$$80 + 60 = 140$$

    c. Show 80 feet less than 125 feet on the number line.

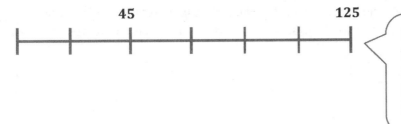

I start by labeling the endpoint on the right. Then I count down by 20's 4 times since it's 80 feet less. Each time, I touch a hash mark on the number line.

    d. Write a subtraction sentence to match the number line.

$$125 - 80 = 45$$

2. Santiago's meter strip got cut off at 49 centimeters.   To measure the length of his eraser, he writes "54 cm — 49 cm." Shirley says it's easier to move the eraser over 1 centimeter. What will Shirley's subtraction sentence be? Explain why she is correct.

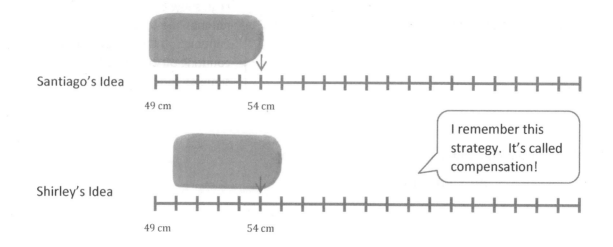

Shirley's subtraction sentence is 55 − 50 = 5. She knows she can move the eraser on the number line, and the length will stay the same. By moving it one unit to the right, she makes an easier problem to solve. 54 − 49 also equals 5, but it's easier to subtract a friendly number like 50 because she only has to subtract the tens.

Lesson 22:   Represent two-digit sums and differences involving length by using the
                        ruler as a number line.

© 2018 Great Minds®. eureka-math.org

Name _____ Date _____

1. Each unit length on both number lines is 10 centimeters.
   (Note: Number lines are not drawn to scale.)

   a. Show 20 centimeters more than 35 centimeters on the number line.

   b. Show 30 centimeters more than 65 centimeters on the number line.

   c. Write an addition sentence to match each number line.

2. Each unit length on both number lines is 5 yards.

   a. Show 35 yards less than 80 yards on the following number line.

   b. Show 25 yards less than 100 yards on the number line.

   c. Write a subtraction sentence to match each number line.

Lesson 22: Represent two-digit sums and differences involving length by using the
           ruler as a number line.

© 2018 Great Minds®. eureka-math.org

179

3. Laura's meter strip got cut off at 37 centimeters. To measure the length of her screwdriver, she writes "50 cm - 37 cm." Tam says it's easier to move the screwdriver over 3 centimeters. What is Tam's subtraction sentence? Explain why she's correct.

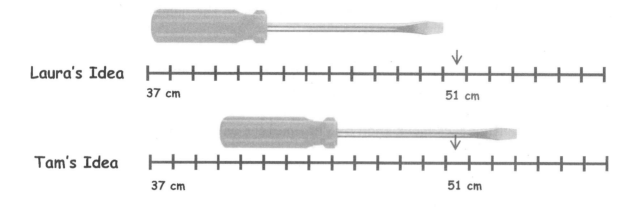

4. Alice measured her belt to be 22 inches long using a yardstick, but she didn't start her measurement at zero. What might be the two endpoints of her belt on her yardstick? Write a subtraction sentence to match your idea.

5. Isaiah ran 100 meters on a 200-meter track. He started running at the 19-meter mark. On what mark did he finish his run?

Lesson 22:   Represent two-digit sums and differences involving length by using the ruler as a number line.

© 2018 Great Minds®. eureka-math.org

1. Measure the length of your shoe and record the length here: **_about 7 inches_**
   Then, measure the length of your family members' shoes, and write the lengths below.

   Name: Shoe Length:

   Mom   **10** _inches_

   Dad   **11** _inches_

   Isaiah (brother)   _about 9 inches_

   Karen (sister)   _about 7 inches_

   > I was very careful to measure everyone's shoe starting at 0 on my ruler.

   > My sister's shoe is a little shorter than 7 inches, and my shoe is a little longer than 7 inches, so both our shoes are about 7 inches.

2. Record your data using tally marks on the table provided.

Shoe Length	Tally of Number of People
Shorter than 9 inches	||
About 9 inches	|
Longer than 9 inches	||

   a. How many more people have a shoe shorter than 9 inches than have a shoe about equal to 9 inches?
   1 person

   b. What is the least common shoe length?
   about 9 inches

   c. Ask and answer one comparison question that can be answered using the data above.

Question: **_How many fewer people have a shoe that is about 9 inches than is longer than 9 inches?_**

Answer: **1 person**

Lesson 23: Collect and record measurement data in a table; answer questions and summarize the data set.

181

© 2019 Great Minds®. eureka-math.org

Name _____   Date _____

Measure your handspan, and record the length here: _____

Then, measure the handspans of your family members, and write the lengths below.

Name: **Handspan:**

_____   _____

_____   _____

_____   _____

_____   _____

_____   _____

1. Record your data using tally marks on the table provided.

Handspan	Tally of Number of People
3 inches	
4 inches	
5 inches	
6 inches	
7 inches	
8 inches	

a. What is the most common handspan length? ____

b. What is the least common handspan length? ____

c. Ask and answer one comparison question that can be answered using the data above.

Question:

_____

_____

Answer:

_____

_____

Lesson 23:   Collect and record measurement data in a table; answer questions and summarize the data set.

183

© 2019 Great Minds®. eureka-math.org

2. a   Use your ruler to measure the lines below in inches. Record the data using tally
        marks on the table provided.

Line A _____

Line B _____

Line C _____

Line D _____

Line E _____

Line F _____

Line G _____

Line Length	Number of Lines
Shorter than 4 inches	
Longer than 4 inches	
Equal to 4 inches	

b. How many more lines are shorter than 4 inches than equal to 4 inches?
    _____

c. What is the difference between the number of lines that are shorter than
    4 inches and those that are longer than 4 inches? _____

d. Ask and answer one comparison question that could be answered using the
    data above.

    Question: _____

    _____

    Answer: _____

    _____

Lesson 23:   Collect and record measurement data in a table; answer questions and
                summarize the data set.

© 2019 Great Minds®. eureka-math.org

EUREKA
MATH

Use the data in the table to create a line plot and answer the questions.

First, I look at the data and count how many pencils there are for each length.

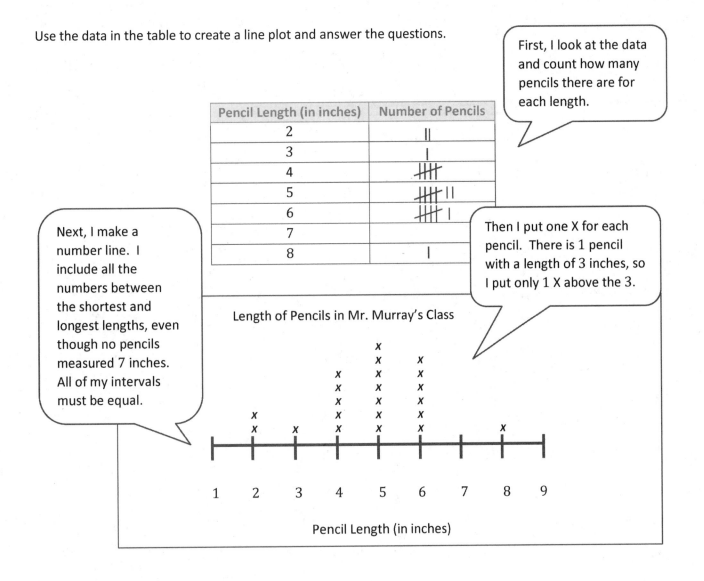

Pencil Length (in inches)	Number of Pencils
2	II
3	I
4	HHT
5	HHT II
6	HHT I
7	
8	I

Next, I make a number line. I include all the numbers between the shortest and longest lengths, even though no pencils measured 7 inches. All of my intervals must be equal.

Then I put one X for each pencil. There is 1 pencil with a length of 3 inches, so I put only 1 X above the 3.

Length of Pencils in Mr. Murray's Class

Pencil Length (in inches)

Describe the pattern you see in the line plot.

The most common pencil length is 5 inches, but 4 inches and 6 inches are also common.

Most of the X's are in the middle of the line plot.

Create your own comparison question related to the data.

How many fewer pencils have a length of 4 inches than a length of 5 inches?

Lesson 24: Draw a line plot to represent the measurement data; relate the measurement scale to the number line.

185

© 2019 Great Minds®. eureka-math.org

Name _____   Date _____

1. Use the data in the table to create a line plot and answer the question.

Handspan (inches)	Number of Students
2	
3	
4	I
5	ℍℍ II
6	ℍℍ ℍℍ
7	III
8	I

Handspans of Students in Ms. DeFransico's Class

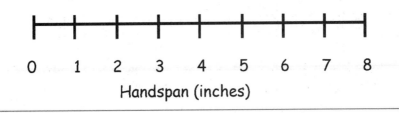

0   1   2   3   4   5   6   7   8

Handspan (inches)

Describe the pattern you see in the line plot:

_____

_____

_____

Lesson 24:   Draw a line plot to represent the measurement data; relate the measurement scale to the number line.

187

© 2019 Great Minds®. eureka-math.org

2. Use the data in the table to create a line plot and answer the questions.

Length of Right Foot (centimeters)	Number of Students
17	I
18	II
19	IIII
20	ⅢⅡ I
21	ⅢⅡ I
22	II
23	I

Lengths of Right Feet of Students in Ms. DeFransico's Class

Line Plot

a. Describe the pattern you see in the line plot.

_____

_____

b. How many feet are longer than 20 centimeters? _____

c. How many feet are shorter than 20 centimeters? _____

d. Create your own comparison question related to the data.

_____

Lesson 24:   Draw a line plot to represent the measurement data; relate the
measurement scale to the number line.

© 2019 Great Minds®. eureka-math.org

Use the data in the table provided to create line plots and answer questions. The table shows the lengths of the daisy chains made at a birthday party.

Length of Daisy Chains	Number of Daisy Chains
3 inches	8
4 inches	5
5 inches	6
7 inches	1
9 inches	3
11 inches	2

> I draw X's above each length to show the data from the table. So I put 8 X's above 3 inches since there are 8 daisy chains that are 3 inches.

> I draw a number line that starts at 3 inches and ends at 11 inches. Since my starting point is 3, I draw a double hash mark to show that the numbers between 0 and 3 are not shown on the scale.

Title ___**Lengths of Daisy Chains**___

```
X
X
X                 X
X        X        X
X        X        X
X        X        X                            X
X        X        X                            X          X
X        X        X              X             X          X
---//---+--------+--------+--------+--------+--------+--------+--------+--------+
0       3        4        5        6        7        8        9        10       11
```

Line Plot   *(in inches)*

> I give my line plot a title, and I label the unit of measure, inches.

a. How many daisy chains were made? ___25___

b. Draw a conclusion about the data in the line plot.

   It is easier to make a short daisy chain. Most of the daisy chains are 5 inches or less.

   Lesson 25:   Draw a line plot to represent a given data set; answer questions and draw conclusions based on measurement data.   **189**

© 2019 Great Minds®. eureka-math.org

c. If 5 more people made 7-inch daisy chains and 6 more people made 9-inch daisy chains, how would it change how the line plot looks?

If 5 more people made 7- inch daisy chains and 6 more people made 9- inch daisy chains, then a 9-inch

chain would be most common, and an 11-inch chain would be least common.

Lesson 25:   Draw a line plot to represent a given data set; answer questions and
                                draw conclusions based on measurement data.

© 2019 Great Minds®. eureka-math.org

Name _____   Date _____

Use the data in the charts provided to create line plots and answer the questions.

1. The chart shows the lengths of the necklaces made in arts and crafts class.

Length of Necklaces	Number of Necklaces
16 inches	3
17 inches	0
18 inches	4
19 inches	0
20 inches	8
21 inches	0
22 inches	9
23 inches	0
24 inches	16

Title _____

Line Plot

a. How many necklaces were made? _____

b. Draw a conclusion about the data in the line plot:

_____

Lesson 25:   Draw a line plot to represent a given data set; answer questions and
draw conclusions based on measurement data.

191

© 2019 Great Minds®. eureka-math.org

2. The chart shows the heights of towers students made with blocks.

Height of Towers	Number of Towers
15 inches	9
16 inches	6
17 inches	2
18 inches	1

Title _____

Line Plot

a. How many towers were measured? _____

b. What tower height occurred most often? _____

c. If 4 more towers were measured at 17 inches and 5 more towers were measured at 18 inches, how would it change how the line plot looks?

_____

_____

d. Draw a conclusion about the data in the line plot:

_____

Lesson 25:   Draw a line plot to represent a given data set; answer questions and draw conclusions based on measurement data.

© 2019 Great Minds®. eureka-math.org

Use the data in the table provided to create a line plot and answer the questions. Plot only the heights of participants given.

The table below describes the heights of pre-schoolers in the soccer game.

Height of Pre-schoolers (in inches)	Number of Pre-schoolers
35	2
37	3
38	6
39	7
40	5
41	2
42	2

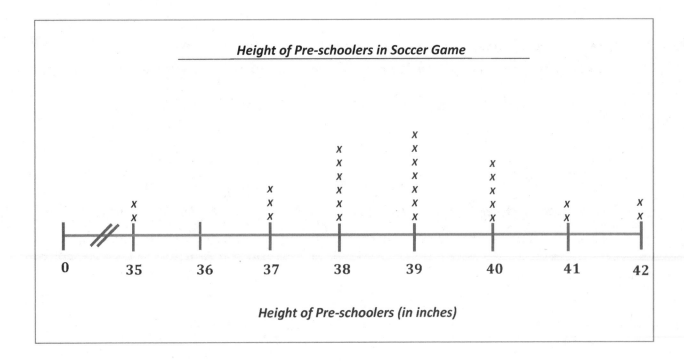

Lesson 26:   Draw a line plot to represent a given data set; answer questions and
             draw conclusions based on measurement data.   193

© 2019 Great Minds®. eureka-math.org

1. How many pre-schoolers were measured? _____27_____

   I started adding with the bigger numbers. I know that $6 + 7 = 13$. Then $13 + 5 = 18$, and 2 more is 20. All that is left is $3 + 2 + 2 = 7$. And $20 + 7 = 27$.

2. How many more pre-schoolers are 38 or 39 inches than 37 or 40 inches? __5__

   I know that 13 pre-schoolers are 38 inches or 39 inches, and 8 pre-schoolers are 37 or 40 inches, so then I just subtract. $13 - 8 = 5$, so the answer is 5 pre-schoolers.

3. Draw a conclusion as to why zero pre-schoolers were between 0 and 35 inches.

   There were 0 pre-schoolers less than 35 inches, because most pre-schoolers are more than 35 inches.

   It would be hard to play on a soccer team if you were only 25 inches tall. That's like a baby!

4. For this data, a **line plot** / **table** (circle one) is easier to read because …

   It is easy to see which heights had the most and least number of pre-schoolers by looking at the number of X's. Also, the measurements are close together, so it's easy to make the number line.

Lesson 26: Draw a line plot to represent a given data set; answer questions and draw conclusions based on measurement data.

© 2019 Great Minds®. eureka-math.org

EUREKA MATH

Name _____   Date _____

Use the data in the table provided to create a line plot and answer the questions. Plot only the lengths of shoelaces given.

1. The table below describes the lengths of student shoelaces in Ms. Henry's class.

Length of Shoelaces (inches)	Number of Shoelaces
27	6
36	10
38	9
40	3
45	2

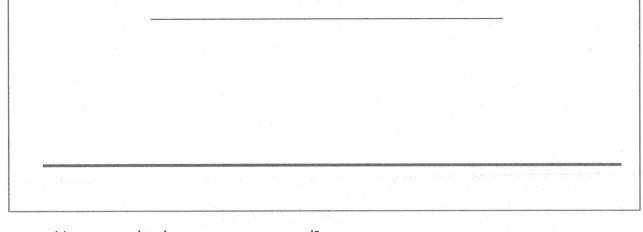

a. How many shoelaces were measured? _____

b. How many more shoelaces are 27 or 36 inches than 40 or 45 inches? _____

c. Draw a conclusion as to why zero students had a 54-inch shoelace.

_____

2. For these data, a **line plot / table** (circle one) is easier to read because...

_____

Lesson 26:   Draw a line plot to represent a given data set; answer questions and   195
             draw conclusions based on measurement data.

© 2019 Great Minds®. eureka-math.org

Use the data in the table provided to create a line plot and answer the questions.

3. The table below describes the lengths of crayons in centimeters in Ms. Harrison's crayon box.

Length (centimeters)	Number of Crayons
4	4
5	7
6	9
7	3
8	1

a. How many crayons are in the box? ___

b. Draw a conclusion as to why most of the crayons are 5 or 6 centimeters:

_____

_____

_____

Lesson 26: Draw a line plot to represent a given data set; answer questions and draw conclusions based on measurement data.

© 2019 Great Minds®. eureka-math.org

EUREKA MATH

Grade 2

Module 8

1. Identify the number of sides and angles for the shape. Circle the angles.

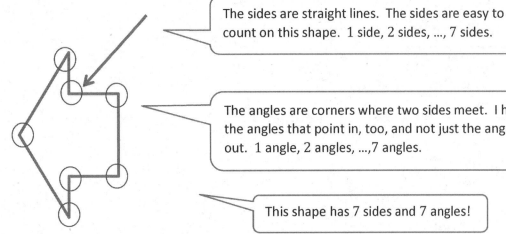

The sides are straight lines. The sides are easy to count on this shape. 1 side, 2 sides, …, 7 sides.

The angles are corners where two sides meet. I have to count the angles that point in, too, and not just the angles that point out. 1 angle, 2 angles, …,7 angles.

This shape has 7 sides and 7 angles!

2. Ethan says that this shape has 6 sides and 6 angles. Frankie says that it has 8 sides and 8 angles. Who is correct? How do you know?

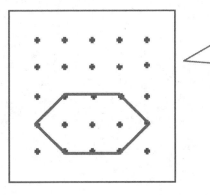

I know that Ethan is correct because I can count 6 sides. I see 3 sides on the top and 3 sides on the bottom. Then I count the angles. I see 3 angles on the left and 3 angles on the right. That means there are 6 sides and 6 angles.

EUREKA MATH®

Lesson 1:   Describe two-dimensional shapes based on attributes.

Name _____   Date _____

1. Identify the number of sides and angles for each shape. Circle each angle as you count, if needed.

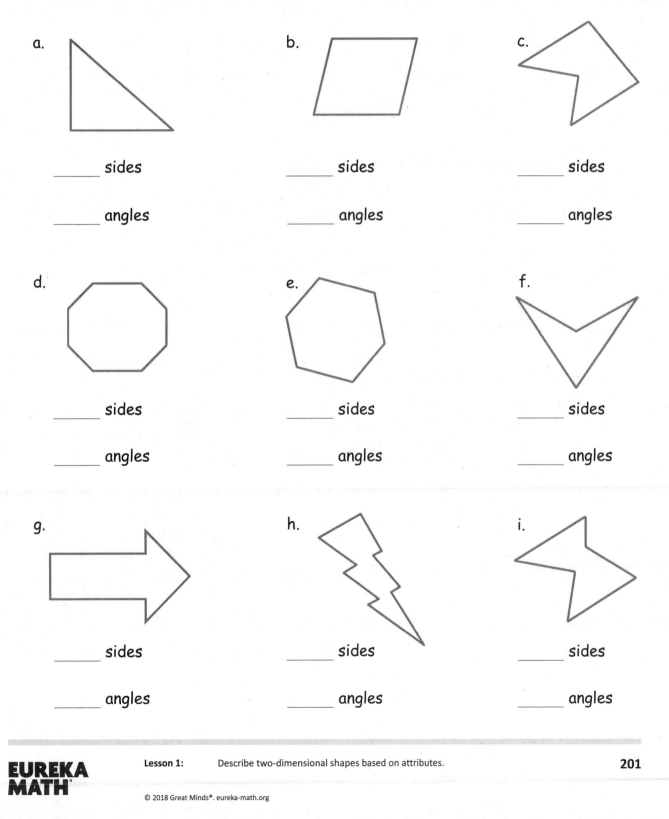

a.

_____ sides

_____ angles

b.

_____ sides

_____ angles

c.

_____ sides

_____ angles

d.

_____ sides

_____ angles

e.

_____ sides

_____ angles

f.

_____ sides

_____ angles

g.

_____ sides

_____ angles

h.

_____ sides

_____ angles

i.

_____ sides

_____ angles

2. Study the shapes below.   Then, answer the questions.

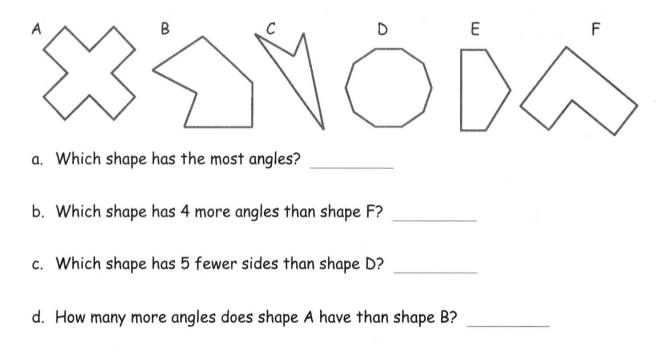

A   B   C   D   E   F

   a. Which shape has the most angles? _____

   b. Which shape has 4 more angles than shape F? _____

   c. Which shape has 5 fewer sides than shape D? _____

   d. How many more angles does shape A have than shape B? _____

   e. Which of these shapes have the same number of sides and angles? _____

3. Joseph's teacher said to make shapes with
   6 sides and 6 angles on his geoboard. Shade
   the shapes that share these attributes, and
   circle the shape that does not belong. Explain
   why it does not belong.

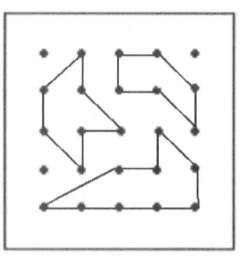

_____

_____

1. Count the number of sides and angles to identify the polygon.

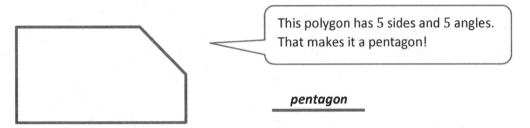

This polygon has 5 sides and 5 angles. That makes it a pentagon!

pentagon

2. Draw more sides to complete 2 examples of the polygon.

	Example 1	Example 2
Pentagon For each example, __3__ lines were added. A pentagon has __5__ total sides.		

3. Explain why both polygons C and D are triangles.

   Both polygons have 3 sides and 3 angles .

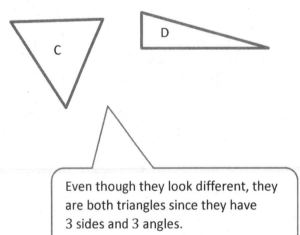

Even though they look different, they are both triangles since they have 3 sides and 3 angles.

Name _____   Date _____

1. Count the number of sides and angles for each shape to identify each polygon. The polygon names in the word bank may be used more than once.

| Hexagon | Quadrilateral | Triangle | Pentagon |

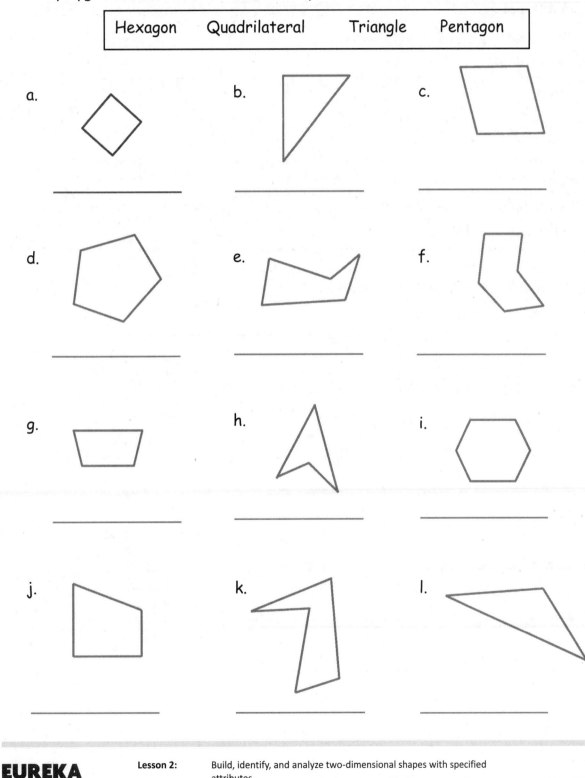

a. _____

b. _____

c. _____

d. _____

e. _____

f. _____

g. _____

h. _____

i. _____

j. _____

k. _____

l. _____

EUREKA
MATH®

Lesson 2: Build, identify, and analyze two-dimensional shapes with specified attributes.

205

© 2018 Great Minds®. eureka-math.org

2. Draw more sides to complete 2 examples of each polygon.

	Example 1	Example 2
a. **Quadrilateral** For each example, ___ lines were added. A quadrilateral has ___ total sides.		
b. **Pentagon** For each example, ___ lines were added. A pentagon has ___ total sides.		
c. **Triangle** For each example, ___ line was added. A triangle has ___ total sides.		
d. **Hexagon** For each example, ___ lines were added. A hexagon has ___ total sides.		

3. Explain why both polygons A and B are pentagons.

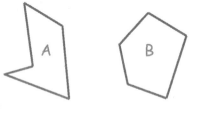

_____

_____

4. Explain why both polygons C and D are triangles.

_____

_____

Lesson 2:   Build, identify, and analyze two-dimensional shapes with specified attributes.

© 2018 Great Minds®. eureka-math.org

1. Use a straightedge to draw the polygon with the given attributes.

   Draw a polygon with 3 angles.

   Number of sides: _____3_____

   Name of polygon: ___*triangle*___

   > When I draw a polygon with 3 angles, it also has 3 sides. That is a triangle!

2. Use your straightedge to draw 2 new examples of the polygon you drew for Problem 1.

   Triangle

   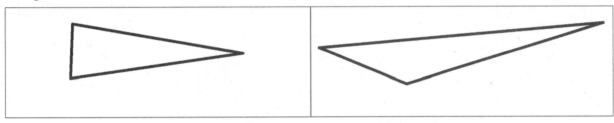

   > All triangles must have 3 sides and 3 angles. By changing the size of the angles and the length of the sides, I can make all kinds of different triangles! This one is long and skinny!

EUREKA MATH®

Lesson 3:   Use attributes to draw different polygons including triangles, quadrilaterals, pentagons, and hexagons.

© 2018 Great Minds®. eureka-math.org

207

Name _____   Date _____

1. Use a straightedge to draw the polygon with the given attributes in the space to the right.

   a. Draw a polygon with 4 angles.

        Number of sides: _____
        Name of polygon: _____

   b. Draw a six-sided polygon.

        Number of angles: _____
        Name of polygon: _____

   c. Draw a polygon with 3 angles.

        Number of sides: _____
        Name of polygon: _____

   d. Draw a five-sided polygon.

        Number of angles: _____
        Name of polygon: _____

EUREKA MATH®

Lesson 3:   Use attributes to draw different polygons including triangles, quadrilaterals, pentagons, and hexagons.

© 2018 Great Minds®. eureka-math.org

209

2. Use your straightedge to draw 2 new examples of each polygon that are different from those you drew on the first page.

a. Quadrilateral

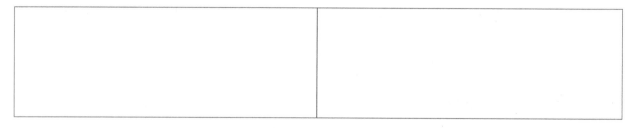

b. Hexagon

c. Pentagon

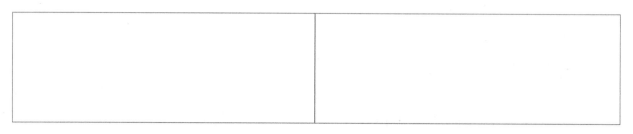

d. Triangle

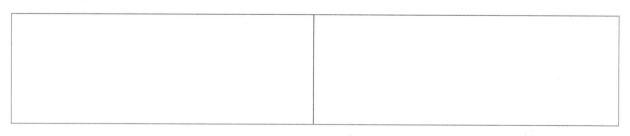

Lesson 3:   Use attributes to draw different polygons including triangles,
                   quadrilaterals, pentagons, and hexagons.

© 2018 Great Minds®. eureka-math.org

1. Use your ruler to draw 2 parallel lines that are not the same length.

_____

I know that parallel lines go in the same direction and never touch. I can draw parallel lines by placing my ruler on the paper and using both sides to draw 2 straight lines.

_____

2. Draw a quadrilateral with 4 square corners.

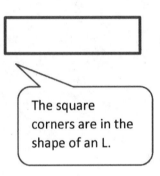

Both of these quadrilaterals have 4 square corners. That means both shapes are rectangles. The one on the right is a special rectangle called a square! It has 4 square corners *and* 4 sides that are the same length!

The square corners are in the shape of an L.

3. Draw a quadrilateral with two sets of parallel sides.

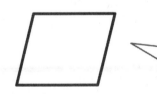

I know this is a quadrilateral because it has 4 sides and 4 angles. It has no square corners, so it can't be a rectangle. It does have 2 sets of parallel sides; it must be a parallelogram!

Lesson 4: Use attributes to identify and draw different quadrilaterals including rectangles, rhombuses, parallelograms, and trapezoids.

211

Name _____   Date _____

1. Use your ruler to draw 2 parallel lines that are not the same length.

2. Use your ruler to draw 2 parallel lines that are the same length.

3. Draw a quadrilateral with two sets of parallel sides. What is the name of this quadrilateral?

4. Draw a quadrilateral with 4 square corners and opposite sides the same length. What is the name of this quadrilateral?

Lesson 4:   Use attributes to identify and draw different quadrilaterals including rectangles, rhombuses, parallelograms, and trapezoids.

© 2018 Great Minds®. eureka-math.org

213

5. A square is a special rectangle. What makes it special?

_____

_____

6. Color each quadrilateral with 4 square corners and two sets of parallel sides red.
    Color each quadrilateral with no square corners and no parallel sides blue.
    Circle each quadrilateral with one or more sets of parallel sides green.

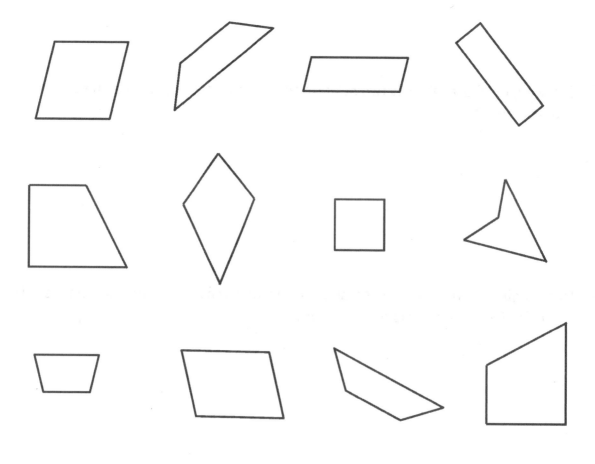

Lesson 4:   Use attributes to identify and draw different quadrilaterals including
rectangles, rhombuses, parallelograms, and trapezoids.

© 2018 Great Minds®. eureka-math.org

Draw a cube.

Step 1:

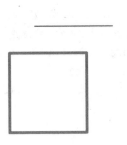

First I draw a square. Then, starting at the middle of the top edge, I draw a line that is parallel to and about the same length as the top edge.

Step 2:

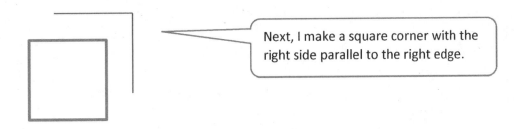

Next, I make a square corner with the right side parallel to the right edge.

Step 3:

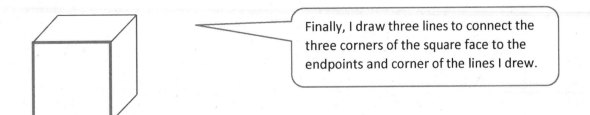

Finally, I draw three lines to connect the three corners of the square face to the endpoints and corner of the lines I drew.

Lesson 5:   Relate the square to the cube, and describe the cube based on attributes.

215

© 2018 Great Minds®. eureka-math.org

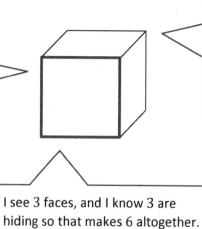

I count the edges by pointing to the ones I see and pointing to the ones I know are hiding! I count 12 edges!

The corners are sharp. There are 4 corners on the front face and 4 corners on the back face. Together that makes 8 corners.

I see 3 faces, and I know 3 are hiding so that makes 6 altogether.

Lesson 5:   Relate the square to the cube, and describe the cube based on attributes.

© 2018 Great Minds®. eureka-math.org

Name _____   Date _____

1. Circle the shapes that could be the face of a cube.

2. What is the most precise name of the shape you circled? _____

3. How many corners does a cube have? _____

4. How many edges does a cube have? _____

5. How many faces does a cube have? _____

6. Draw 6 cubes, and put a star next to your best one.

First cube	Second cube
Third cube	Fourth cube
Fifth cube	Sixth cube

Lesson 5:   Relate the square to the cube, and describe the cube based on attributes.

© 2018 Great Minds®. eureka-math.org

217

7. Connect the corners of the squares to make a different kind of drawing of a cube.

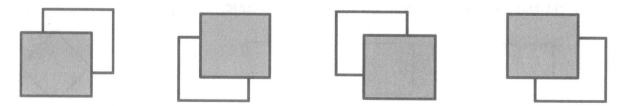

8. Patricia used the image of the cube below to count 7 corners. Explain where the 8th corner is hiding.

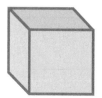

_____

_____

_____

218

Lesson 5:   Relate the square to the cube, and describe the cube based on attributes.

© 2018 Great Minds®. eureka-math.org

1. Identify each polygon labeled in the tangram as precisely as possible in the space below.

a. _____ *triangle* _____

b. _____ *parallelogram* _____

c. _____ *square* _____

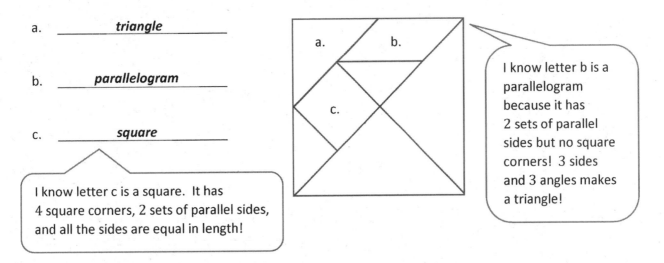

I know letter b is a parallelogram because it has 2 sets of parallel sides but no square corners! 3 sides and 3 angles makes a triangle!

I know letter c is a square. It has 4 square corners, 2 sets of parallel sides, and all the sides are equal in length!

2. Use the parallelogram and the two smallest triangles to make the following polygons. Draw them in the space provided.

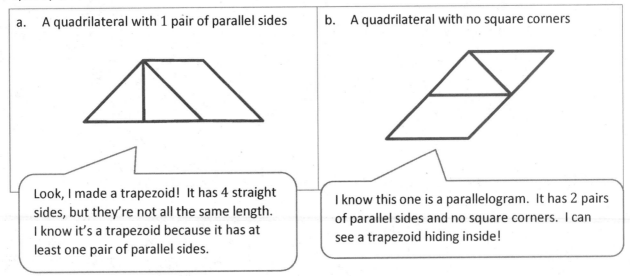

a. A quadrilateral with 1 pair of parallel sides

b. A quadrilateral with no square corners

Look, I made a trapezoid! It has 4 straight sides, but they're not all the same length. I know it's a trapezoid because it has at least one pair of parallel sides.

I know this one is a parallelogram. It has 2 pairs of parallel sides and no square corners. I can see a trapezoid hiding inside!

Lesson 6:   Combine shapes to create a composite shape; create a new shape from composite shapes.

219

Name _____   Date _____

1. Identify each polygon labeled in the tangram as precisely as possible in the space below.

   a. _____

   b. _____

   c. _____

2. Use the square and the two smallest triangles of your tangram piecees to make the following polygons. Draw them in the space provided.

a. A triangle with 1 square corner.	b. A quadrilateral with 4 square corners.
c. A quadrilateral with no square corners.	d. A quadrilateral with only 1 pair of parallel sides.

Lesson 6:   Combine shapes to create a composite shape; create a new shape from composite shapes.

221

© 2018 Great Minds®. eureka-math.org

3. Rearrange the parallelogram and the two smallest triangles of your tangram pieces to make a hexagon. Draw the new shape below.

4. Rearrange your tangram pieces to make at least 6 other polygons! Draw and name them below.

Lesson 6:   Combine shapes to create a composite shape; create a new shape
               from composite shapes.

Cut the tangram into 7 puzzle pieces.

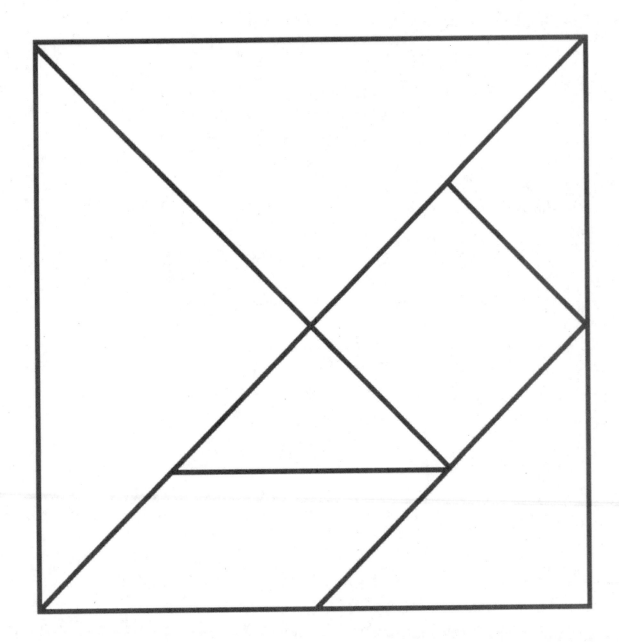

tangram

Lesson 6:   Combine shapes to create a composite shape; create a new shape
             from composite shapes.

© 2018 Great Minds®. eureka-math.org

223

1. Solve the following puzzle using your tangram pieces. Draw your solutions in the space below.

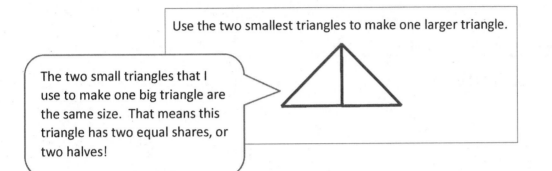

Use the two smallest triangles to make one larger triangle.

The two small triangles that I use to make one big triangle are the same size. That means this triangle has two equal shares, or two halves!

2. Circle the shapes that show thirds.

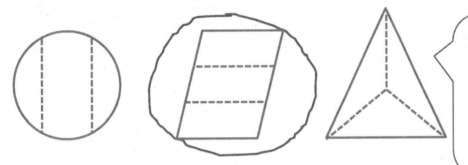

I know this triangle is not cut into thirds because all three parts are not equal shares. The bottom part is bigger than the other ones!

3. Examine the rectangle.

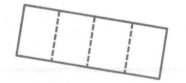

a. How many equal shares does the rectangle have? ___4___

b. How many fourths are in the rectangle? ___4___

Lesson 7:   Interpret equal shares in composite shapes as halves, thirds, and fourths.

225

© 2018 Great Minds®. eureka-math.org

Name _____ Date _____

1. Solve the following puzzles using your tangram pieces. Draw your solutions in the space below.

a. Use the two largest triangles to make a square.	b. Use the two smallest triangles to make a square.
c. Use the two smallest triangles to make a parallelogram with no square corners.	d. Use the two smallest triangles to make one larger triangle.
e. How many equal shares do the larger shapes in Parts (a–d) have?	f. How many halves make up the larger shapes in Parts (a–d)?

2. Circle the shapes that show halves.

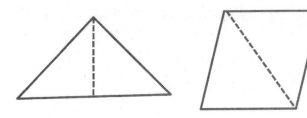 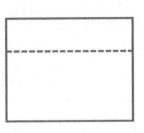 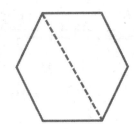

Lesson 7:   Interpret equal shares in composite shapes as halves, thirds, and fourths.

227

© 2018 Great Minds®. eureka-math.org

3. Examine the trapezoid.

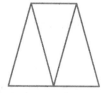

   a. How many equal shares does the trapezoid have? _____

   b. How many thirds are in the trapezoid? _____

4. Circle the shapes that show thirds.

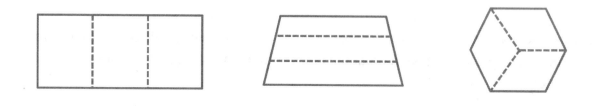

5. Examine the parallelogram.

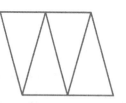

   a. How many equal shares does the shape have? _____

   b. How many fourths are in the shape? _____

6. Circle the shapes that show fourths.

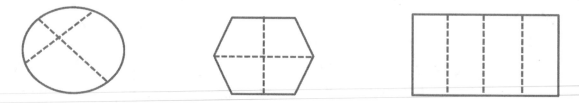

Lesson 7:   Interpret equal shares in composite shapes as halves, thirds, and
             fourths.

EUREKA
MATH

1. Name the pattern block used to cover half the rectangle. _____**square**_____

    Sketch the 2 pattern blocks used to cover both halves of the rectangle.

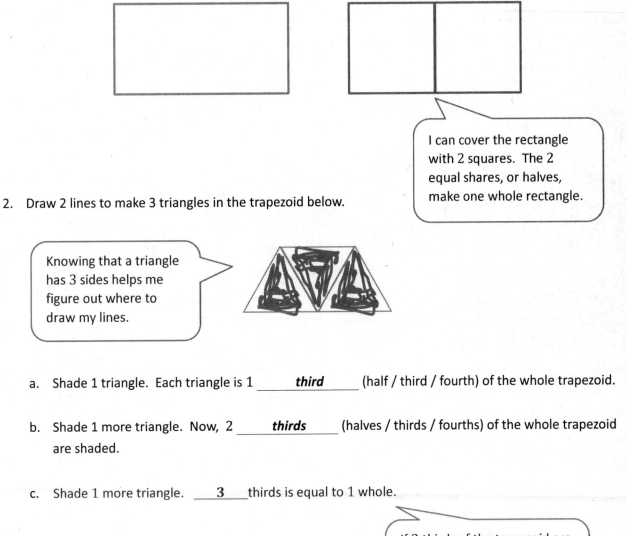

> I can cover the rectangle with 2 squares. The 2 equal shares, or halves, make one whole rectangle.

2. Draw 2 lines to make 3 triangles in the trapezoid below.

> Knowing that a triangle has 3 sides helps me figure out where to draw my lines.

a. Shade 1 triangle. Each triangle is 1 _____**third**_____ (half / third / fourth) of the whole trapezoid.

b. Shade 1 more triangle. Now, 2 _____**thirds**_____ (halves / thirds / fourths) of the whole trapezoid are shaded.

c. Shade 1 more triangle. _____**3**_____ thirds is equal to 1 whole.

> If 2 thirds of the trapezoid are shaded, I have 1 third left to shade. Then, 3 thirds will be shaded. That's 1 whole!

Lesson 8:   Interpret equal shares in composite shapes as halves, thirds, and fourths.

229

© 2018 Great Minds®. eureka-math.org

Name _____   Date _____

1. Name the pattern block used to cover half the rhombus. _____

   Sketch the 2 pattern blocks used to cover both halves of the rhombus.

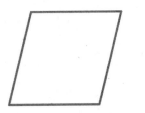

2. Name the pattern block used to cover half the hexagon. _____

   Sketch the 2 pattern blocks used to cover both halves of the hexagon.

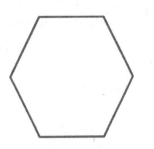

3. Name the pattern block used to cover 1 third of the hexagon. _____

   Sketch the 3 pattern blocks used to cover thirds of the hexagon.

4. Name the pattern block used to cover 1 third of the trapezoid. _____

   Sketch the 3 pattern blocks used to cover thirds of the trapezoid.

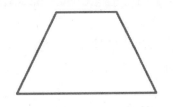

Lesson 8:   Interpret equal shares in composite shapes as halves, thirds, and fourths.

© 2018 Great Minds®. eureka-math.org

231

5. Draw 2 lines to make 4 squares in the square below.

    a. Shade 1 small square. Each small square is 1 _____ (half / third / fourth) of the whole square.

    b. Shade 1 more small square. Now, 2 _____ (halves / thirds / fourths) of the whole square are shaded.

    c. And 2 fourths of the square is the same as 1 _____ (half / third / fourth) of the whole square.

    d. Shade 2 more small squares. _____fourths is equal to 1 whole.

6. Name the pattern block used to cover 1 sixth of the hexagon. _____
   Sketch the 6 pattern blocks used to cover 6 sixths of the hexagon.

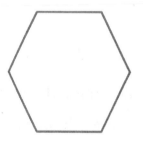

Lesson 8:   Interpret equal shares in composite shapes as halves, thirds, and fourths.

© 2018 Great Minds®. eureka-math.org

EUREKA
MATH

1. Circle the shapes that have 2 equal shares with 1 share shaded.

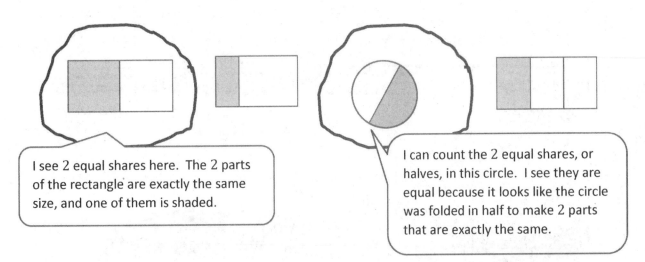

I see 2 equal shares here. The 2 parts of the rectangle are exactly the same size, and one of them is shaded.

I can count the 2 equal shares, or halves, in this circle. I see they are equal because it looks like the circle was folded in half to make 2 parts that are exactly the same.

2. Shade 1 half of the shapes that are split into 2 equal shares. One has been done for you.

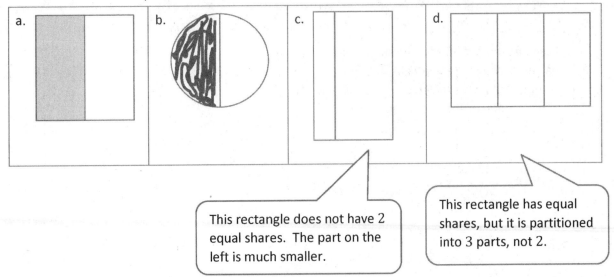

This rectangle does not have 2 equal shares. The part on the left is much smaller.

This rectangle has equal shares, but it is partitioned into 3 parts, not 2.

Lesson 9:   Partition circles and rectangles into equal parts, and describe those parts as halves, thirds, or fourths.

© 2018 Great Minds®. eureka-math.org

233

3. Partition the shapes to show halves. Shade 1 half of each. Compare your halves to your partner's.

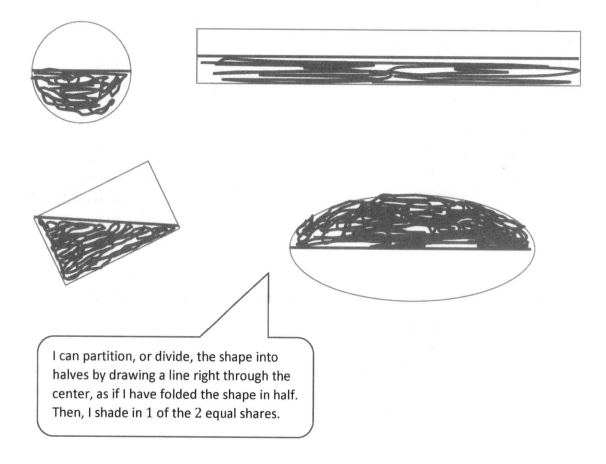

I can partition, or divide, the shape into halves by drawing a line right through the center, as if I have folded the shape in half. Then, I shade in 1 of the 2 equal shares.

Lesson 9:   Partition circles and rectangles into equal parts, and describe those
               parts as halves, thirds, or fourths.

Name _____   Date _____

1. Circle the shapes that have 2 equal shares with 1 share shaded.

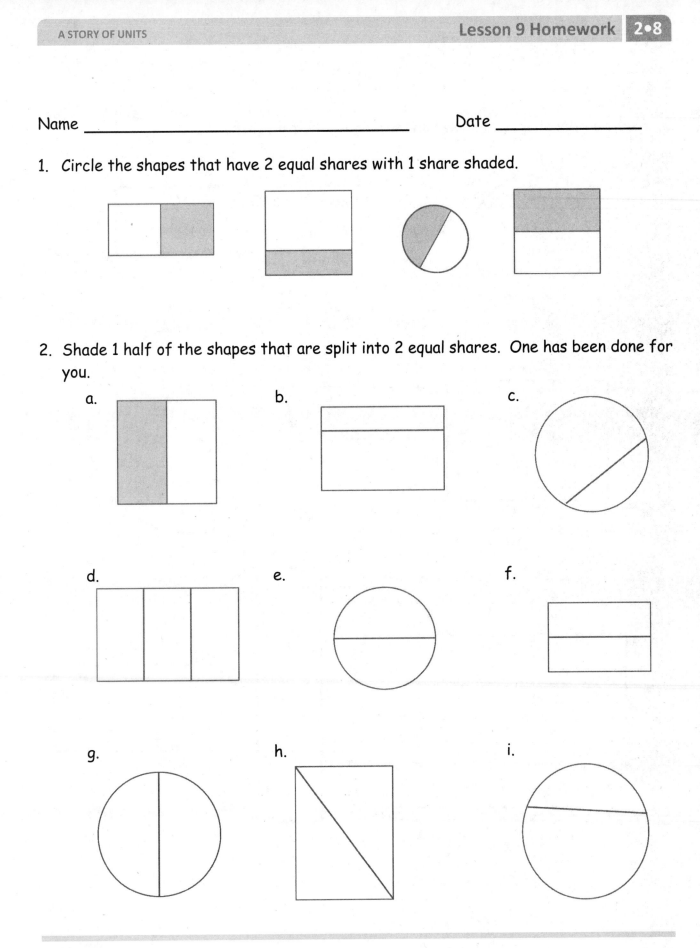

2. Shade 1 half of the shapes that are split into 2 equal shares. One has been done for you.

a.

b.

c.

d.

e.

f.

g.

h.

i.

EUREKA
MATH®

Lesson 9:   Partition circles and rectangles into equal parts, and describe those
parts as halves, thirds, or fourths.

235

© 2018 Great Minds®. eureka-math.org

3. Partition the shapes to show halves. Shade 1 half of each.

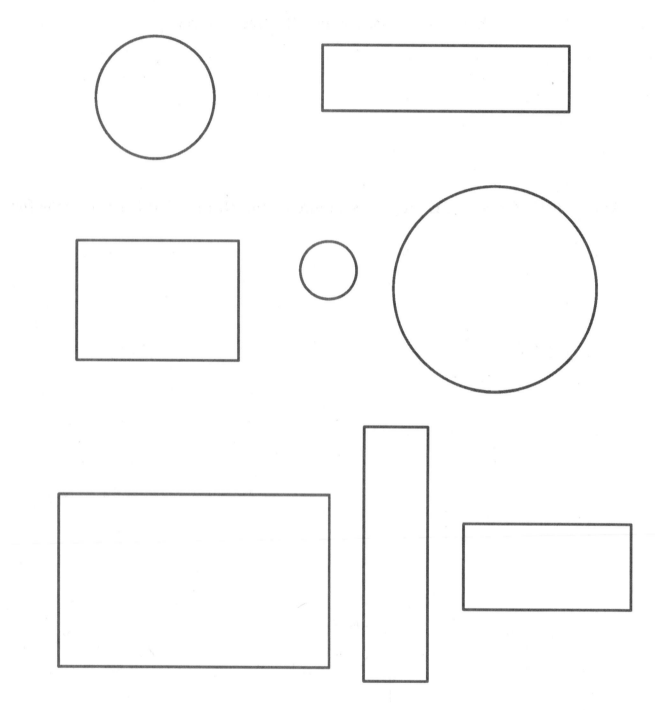

Lesson 9:   Partition circles and rectangles into equal parts, and describe those
            parts as halves, thirds, or fourths.

© 2018 Great Minds®. eureka-math.org

EUREKA
MATH®

I know that these shapes show halves because each shape has 2 equal shares.

1. Do the shapes below show halves or thirds? _____halves_____

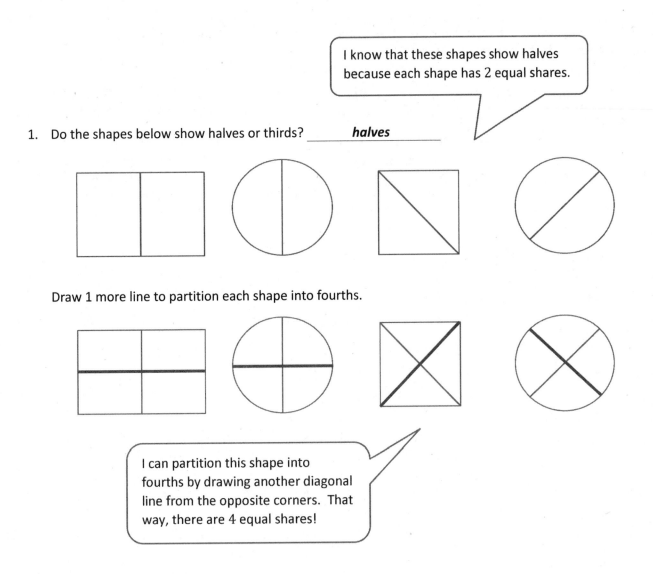

Draw 1 more line to partition each shape into fourths.

I can partition this shape into fourths by drawing another diagonal line from the opposite corners. That way, there are 4 equal shares!

2. Partition each rectangle into fourths. Then, shade the shapes as indicated.

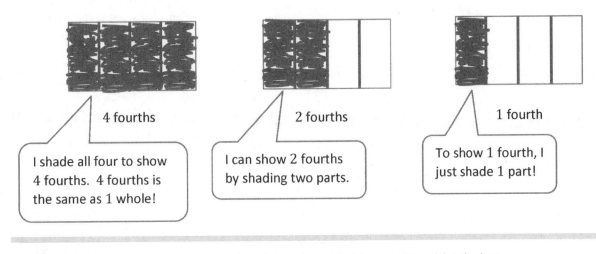

4 fourths

2 fourths

1 fourth

I shade all four to show 4 fourths. 4 fourths is the same as 1 whole!

I can show 2 fourths by shading two parts.

To show 1 fourth, I just shade 1 part!

Lesson 10:   Partition circles and rectangles into equal parts, and describe those
parts as halves, thirds, or fourths.

237

© 2018 Great Minds®. eureka-math.org

3. Split the granola bar below so that Lisa, MJ, and Jessa all have an equal share.
    Label each student's share with her name.

Lisa	MJ	Jessa

    What fraction of the granola bar did the girls get in all?

    3 thirds

    They shared the whole granola bar! That is 3 thirds!

    I split the bar into 3 equal shares because there are 3 people eating it!

Lesson 10:   Partition circles and rectangles into equal parts, and describe those
                        parts as halves, thirds, or fourths.

© 2018 Great Minds®. eureka-math.org

Name _____ Date _____

1. a. Do the shapes below show halves or thirds? _____

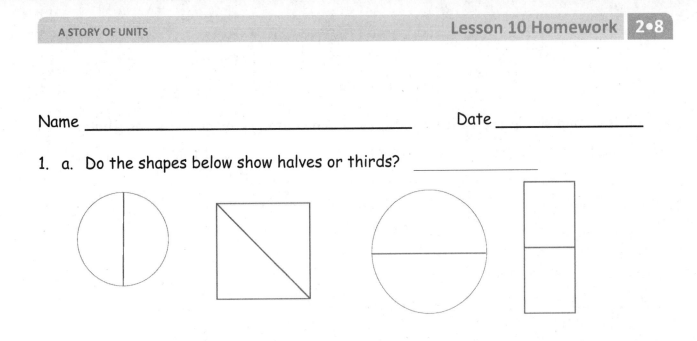

   b. Draw 1 more line to partition each shape above into fourths.

2. Partition each rectangle into thirds. Then, shade the shapes as indicated.

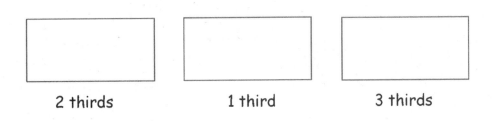

   2 thirds   1 third   3 thirds

3. Partition each circle into fourths. Then, shade the shapes as indicated.

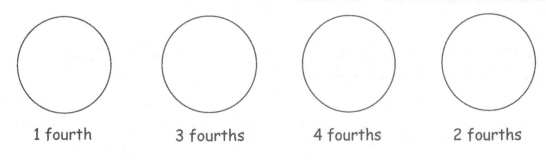

   1 fourth   3 fourths   4 fourths   2 fourths

Lesson 10: Partition circles and rectangles into equal parts, and describe those parts as halves, thirds, or fourths.

© 2018 Great Minds®. eureka-math.org

239

4. Partition and shade the following shapes. Each rectangle or circle is one whole.

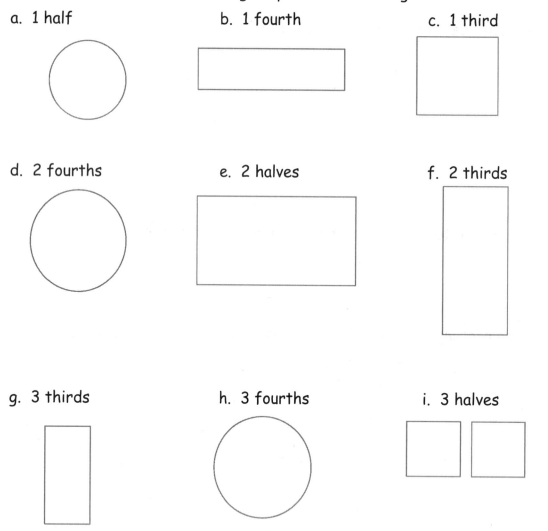

   a. 1 half   b. 1 fourth   c. 1 third

   d. 2 fourths   e. 2 halves   f. 2 thirds

   g. 3 thirds   h. 3 fourths   i. 3 halves

5. Split the pizza below so that Shane, Raul, and John all have an equal share.
   Label each student's share with his name.

   What fraction of the pizza did the boys get in all?

Lesson 10:   Partition circles and rectangles into equal parts, and describe those
                                parts as halves, thirds, or fourths.   **EUREKA
                  © 2018 Great Minds®. eureka-math.org   MATH**

1. For part (a), identify the shaded area.

   a.

   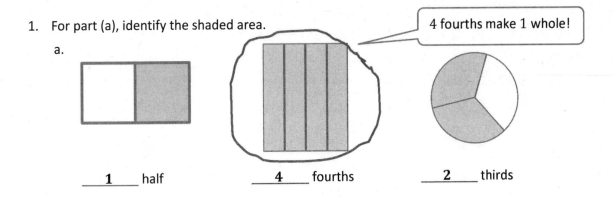

   4 fourths make 1 whole!

   ___1___ half   ___4___ fourths   ___2___ thirds

   b. Circle the shape above that has a shaded area that shows 1 whole.

2. What fraction do you need to color so that 1 whole is shaded?

   a.   b.

   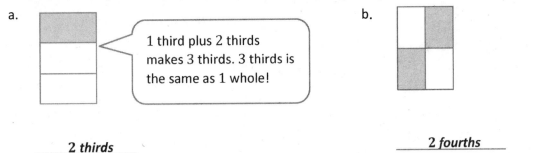

   1 third plus 2 thirds makes 3 thirds. 3 thirds is the same as 1 whole!

   _____2 thirds_____   _____2 fourths_____

3. Complete the drawing to show 1 whole.

   This is 1 third.

   Draw 1 whole.

   1 third and 1 third and 1 third make a whole; 3 thirds make a whole. I only have 1 third, so I need to draw 2 more!

EUREKA MATH®   Lesson 11:   Describe a whole by the number of equal parts including 2 halves, 3   **241**
thirds, and 4 fourths.

© 2018 Great Minds®. eureka-math.org

Name _____ Date _____

1. For Parts (a), (c), and (e), identify the shaded area.

   a.

   _____ half   _____ halves

   b. Circle the shape above that has a shaded area that shows 1 whole.

   c.

   _____ third   _____ thirds   _____ thirds

   d. Circle the shape above that has a shaded area that shows 1 whole.

   e.

   _____ fourth   _____ fourths   _____ fourths   _____ fourths

   f. Circle the shape above that has a shaded area that shows 1 whole.

Lesson 11:   Describe a whole by the number of equal parts including 2 halves, 3 thirds, and 4 fourths.

© 2018 Great Minds®. eureka-math.org

243

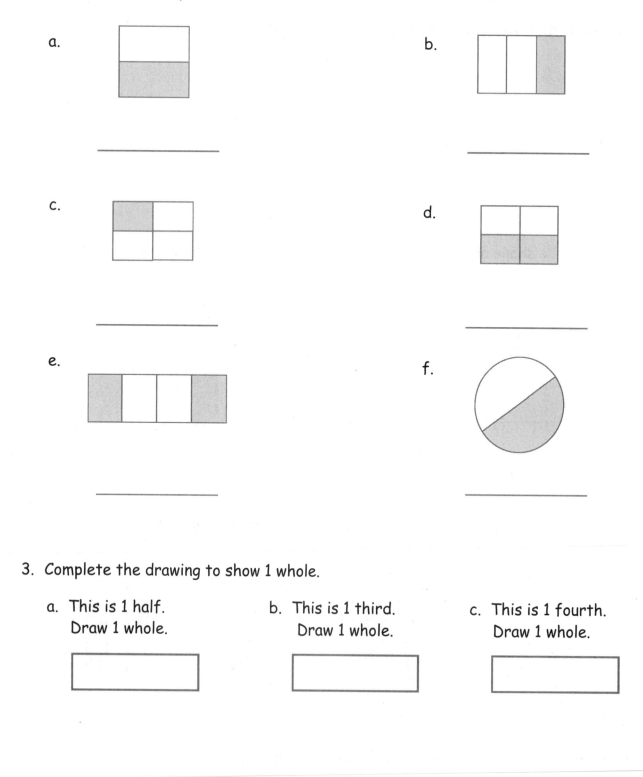

2. What fraction do you need to color so that 1 whole is shaded?

a.

_____

b.

_____

c.

_____

d.

_____

e.

_____

f.

_____

3. Complete the drawing to show 1 whole.

   a. This is 1 half.
      Draw 1 whole.

   b. This is 1 third.
      Draw 1 whole.

   c. This is 1 fourth.
      Draw 1 whole.

Lesson 11:   Describe a whole by the number of equal parts including 2 halves, 3 thirds, and 4 fourths.

© 2018 Great Minds®. eureka-math.org

EUREKA
MATH®

1. Partition the rectangles in 2 different ways to show equal shares.

    2 halves

    Look, I can show thirds as long, skinny rectangles or short, fat rectangles!
    They don't need to have the same shape to cover the same amount of space.

    3 thirds

    I can show fourths in more than one way! As long as the 4 parts cover
    the same amount of space they are equal, so I have made fourths!

    4 fourths

Lesson 12: Recognize that equal parts of an identical rectangle can have different shapes.

245

© 2018 Great Minds®. eureka-math.org

2. Cut out the rectangle.

a. Cut the rectangle in half to make 2 equal size rectangles. Shade 1 half using your pencil.

I can make 2 equal size rectangles by folding my paper in half the long way.

b. Rearrange the halves to create a new rectangle with no gaps or overlaps.

I can line up the rectangles with no gaps or overlaps by making the ends touch.

c. Cut each equal part in half to make 4 equal size rectangles.

I have 2 equal rectangles. If I cut each rectangle into 2 equal shares, I will have 4 equal size rectangles! Now 2 fourths are shaded.

d. Rearrange the new equal shares to create different polygons.

e. Draw one of your new polygons from part (d) below. One half is shaded!

Even though I have a shape that looks different, one half is still shaded!

Lesson 12: Recognize that equal parts of an identical rectangle can have different shapes.

© 2018 Great Minds®. eureka-math.org

Name _____   Date _____

1. Partition the rectangles in 2 different ways to show equal shares.

   a. 2 halves

   b. 3 thirds

   c. 4 fourths

   d. 2 halves

   e. 3 thirds

   f. 4 fourths

EUREKA MATH

Lesson 12:   Recognize that equal parts of an identical rectangle can have different shapes.

© 2018 Great Minds®. eureka-math.org

247

2. Cut out the square at the bottom of this page.

   a. Cut the square in half to make 2 equal-size rectangles. Shade 1 half using your pencil.

   b. Rearrange the halves to create a new rectangle with no gaps or overlaps.

   c. Cut each equal part in half to make 4 equal-size squares.

   d. Rearrange the new equal shares to create different polygons.

   e. Draw one of your new polygons from Part (d) below. One half is shaded!

Lesson 12:   Recognize that equal parts of an identical rectangle can have different
                                           shapes.

© 2018 Great Minds®. eureka-math.org

EUREKA
MATH

1. Tell what fraction of each clock is shaded in the space below using the words *quarter, quarters, half,* or *halves.*

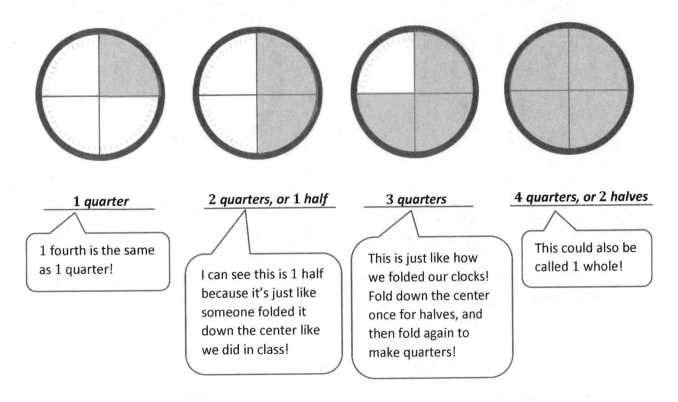

1 quarter

1 fourth is the same as 1 quarter!

2 quarters, or 1 half

I can see this is 1 half because it's just like someone folded it down the center like we did in class!

3 quarters

This is just like how we folded our clocks! Fold down the center once for halves, and then fold again to make quarters!

4 quarters, or 2 halves

This could also be called 1 whole!

Lesson 13: Construct a paper clock by partitioning a circle into halves and quarters, and tell time to the half hour or quarter hour.

249

© 2018 Great Minds®. eureka-math.org

2. Write the time shown on each clock.

a.

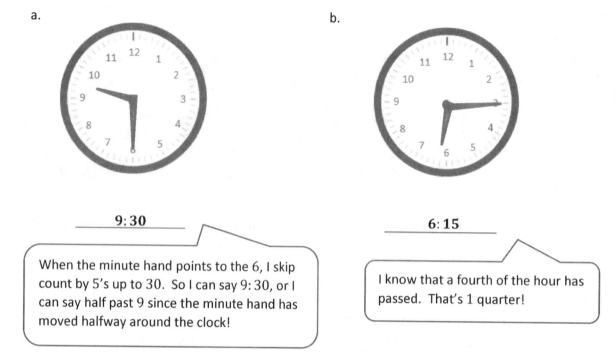

9:30

When the minute hand points to the 6, I skip count by 5's up to 30. So I can say 9:30, or I can say half past 9 since the minute hand has moved halfway around the clock!

b.

6:15

I know that a fourth of the hour has passed. That's 1 quarter!

3. Draw the minute hand on the clock to show the correct time.

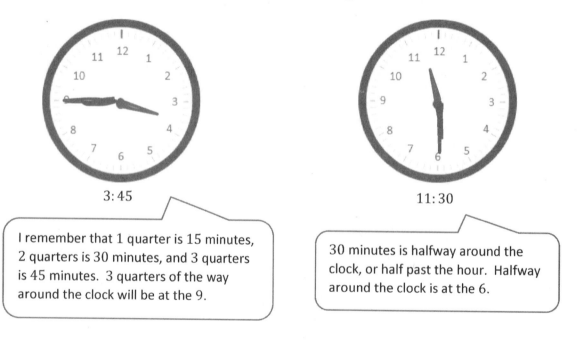

3:45

I remember that 1 quarter is 15 minutes, 2 quarters is 30 minutes, and 3 quarters is 45 minutes. 3 quarters of the way around the clock will be at the 9.

11:30

30 minutes is halfway around the clock, or half past the hour. Halfway around the clock is at the 6.

Lesson 13:   Construct a paper clock by partitioning a circle into halves and quarters, and tell time to the half hour or quarter hour.

© 2018 Great Minds®. eureka-math.org

Name _____   Date _____

1. Tell what fraction of each clock is shaded in the space below using the words *quarter, quarters, half,* or *halves.*

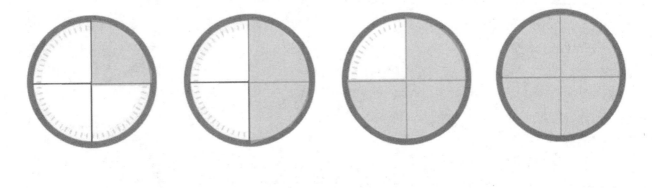

_____   _____   _____   _____

2. Write the time shown on each clock.

a.

b.

_____   _____

c.

d.

_____   _____

Lesson 13:   Construct a paper clock by partitioning a circle into halves and
             quarters, and tell time to the half hour or quarter hour.   251

© 2018 Great Minds®. eureka-math.org

3. Match each time to the correct clock by drawing a line.

- Quarter to 5

- Half past 5

- 5:15

- Quarter after 5

- 4:45

4. Draw the minute hand on the clock to show the correct time.

3:30   11:45   6:15

Lesson 13:   Construct a paper clock by partitioning a circle into halves and quarters, and tell time to the half hour or quarter hour.

1. Fill in the missing numbers.

60, 55, 50, __45__, 40, __35__, __30__, __25__, __20__, __15__, __10__, __5__, __0__

> I skip-count back by 5's. It's just like counting back around the clock!

2. Draw the hour and minute hands on the clocks to match the correct time.

3:05

> I know that since it is only 5 minutes past the hour, the hour hand should be pointing at the 3.

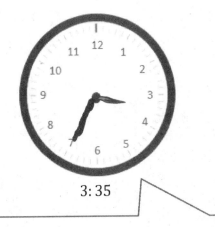

3:35

> More than half of the hour has passed, so the hour hand should be pointing about halfway between the 3 and 4. I know that when the minute hand is pointing to the 6, it is 30 minutes past the hour. When it's pointing to the 7, I add on 5 minutes, so the clock shows 3:35.

6:55

> Since it's 6:55, that means it is almost 7. The hour hand should be pointing right before the 7 since it's just 5 minutes before 7 o'clock.

Name _____   Date _____

1. Fill in the missing numbers.

    0, 5, 10, _____, _____, _____, _____, 35, _____, _____, _____, _____, _____

    _____, _____, _____, 45, 40, _____, _____, _____, 20, 15, _____, _____, _____

2. Fill in the missing minutes on the face of the clock.

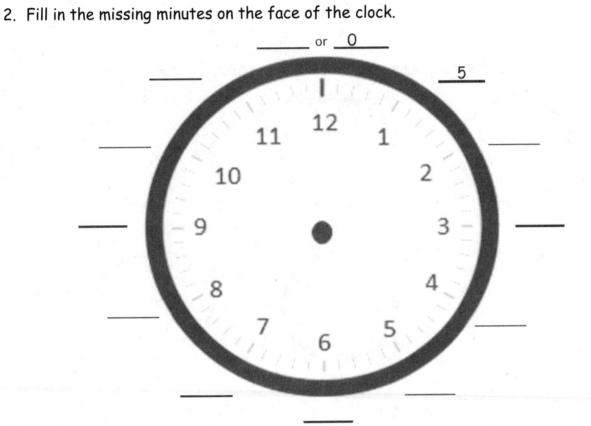

3. Draw the minute hands on the clocks to match the correct time.

3:25   7:15   9:55

4. Draw the hour hands on the clocks to match the correct time.

12:30   10:10   3:45

5. Draw the hour and minute hands on the clocks to match the correct time.

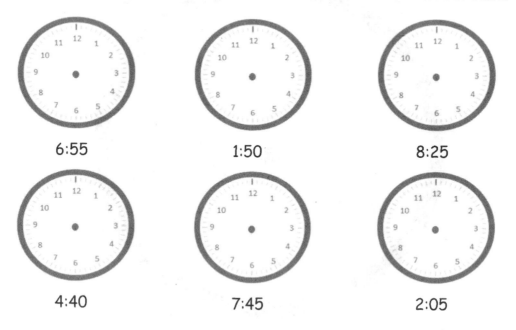

6:55   1:50   8:25

4:40   7:45   2:05

6. What time is it?

_____   _____

© 2018 Great Minds®. eureka-math.org

1. Decide whether the activity below would happen in the a.m. or the p.m. Circle your answer.

   Waking up for school   (a.m.)/ p.m.
   Eating dinner   a.m. /(p.m.)
   Reading a bedtime story   a.m. /(p.m.)
   Making breakfast   (a.m.)/ p.m.

   > *A comes before P in the alphabet. That's how I remember that a.m. is morning and p.m. is afternoon. The morning comes before the afternoon!*

2. What time does the clock show?

   __3__ : __55__

   > *Even though it looks like the hour hand is pointing to the 4, I know it's not 4 o'clock yet because the minute hand shows 55 minutes! I have to wait 5 more minutes!*

3. Draw the hands on the analog clock to match the time on the digital clock. Then, circle a.m. or p.m. based on the description given.

   Brushing your teeth after you wake up

   7 : 10   (a.m.) or p.m.

   > *I know it's a.m. because it says "after you wake up," and that happens in the morning!*

   > *The digital time shows the digits of the hour and the minutes. On the analog clock, the little hand points to the 7 to show the hour. For the minute hand, I can count by 5's to figure out how to show 10 minutes after the hour. 5, 10…so the big hand points to the 2 to show 10 minutes.*

4. Write what you might be doing if it were a.m. or p.m.

   a.m. _____*eating breakfast*_____
   p.m. _____*reading a book*_____

   > *Usually at 7 in the morning, I am eating breakfast. 7 p.m. is 1 hour before bed, and that's the time I read!*

Lesson 15:   Tell time to the nearest five minutes; relate *a.m.* and *p.m.* to time of day.

© 2018 Great Minds®. eureka-math.org

257

Name _____ Date _____

1. Decide whether the activity below would happen in the a.m. or the p.m. Circle your answer.

a. Eating breakfast	a.m. / p.m.	b. Doing homework	a.m. / p.m.
c. Setting the table for dinner	a.m. / p.m.	d. Waking up in the morning	a.m. / p.m.
e. After-school dance class	a.m. / p.m.	f. Eating lunch	a.m. / p.m.
g. Going to bed	a.m. / p.m.	h. Heating up dinner	a.m. / p.m.

2. Write the time displayed on the clock. Then, choose whether the activity below would happen in the a.m. or the p.m.

a. Brushing your teeth before school	b. Eating dessert after dinner
_____ : _____ a.m. / p.m.	_____ : _____ a.m. / p.m.

Lesson 15: Tell time to the nearest five minutes; relate *a.m.* and *p.m.* to time of day.

259

© 2018 Great Minds®. eureka-math.org

3. Draw the hands on the analog clock to match the time on the digital clock. Then, circle **a.m.** or **p.m.** based on the description given.

   a. Brushing your teeth before bedtime

    a.m. or p.m.

   b. Recess after lunch

   12:30   a.m. or p.m.

4. Write what you might be doing if it were **a.m.** or **p.m.**

    a. **a.m.** _____

    b. **p.m.** _____

    c. **a.m.** _____

    d. **p.m.** _____

Lesson 15:   Tell time to the nearest five minutes; relate *a.m.* and *p.m.* to time of day.

© 2018 Great Minds®. eureka-math.org

1   How much time has passed?

6: 30 a.m. → 7: 00 a.m.   __30 *minutes*__

> 6: 30 is half past the hour. That means that it takes another half to get to the next hour, so 30 minutes have passed.

4: 00 p.m. → 9: 00 p.m.   __5 *hours*__

> I can add on from 4: 00 p.m. to get to 9: 00 p.m. $4 + 5 = 9$, so 5 hours have passed.

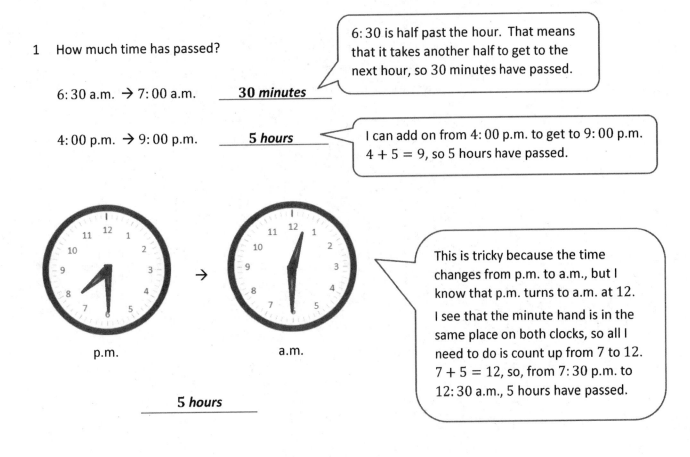

p.m.   a.m.

> This is tricky because the time changes from p.m. to a.m., but I know that p.m. turns to a.m. at 12.
>
> I see that the minute hand is in the same place on both clocks, so all I need to do is count up from 7 to 12. $7 + 5 = 12$, so, from 7: 30 p.m. to 12: 30 a.m., 5 hours have passed.

__5 *hours*__

2.   Anna spent 3 hours at dance practice. She finished at 6:15 p.m. What time did she start?

?   ——— **+ 3 hours** ——→   6: 15

> I can use the arrow way with hours and minutes to make solving easier.

$6 - 3 = 3$, so 6: 15 *minus 3 hours is* 3: 15.

Anna started at 3: 15.

Name _____   Date _____

1. How much time has passed?

   a. 2:00 p.m. → 8:00 p.m.   _____

   b. 7:30 a.m. → 12:00 p.m. (noon) _____

   c. 10:00 a.m. → 4:30 p.m.   _____

   d. 1:30 p.m. → 8:30 p.m.   _____

   e. 9:30 a.m. → 2:00 p.m.   _____

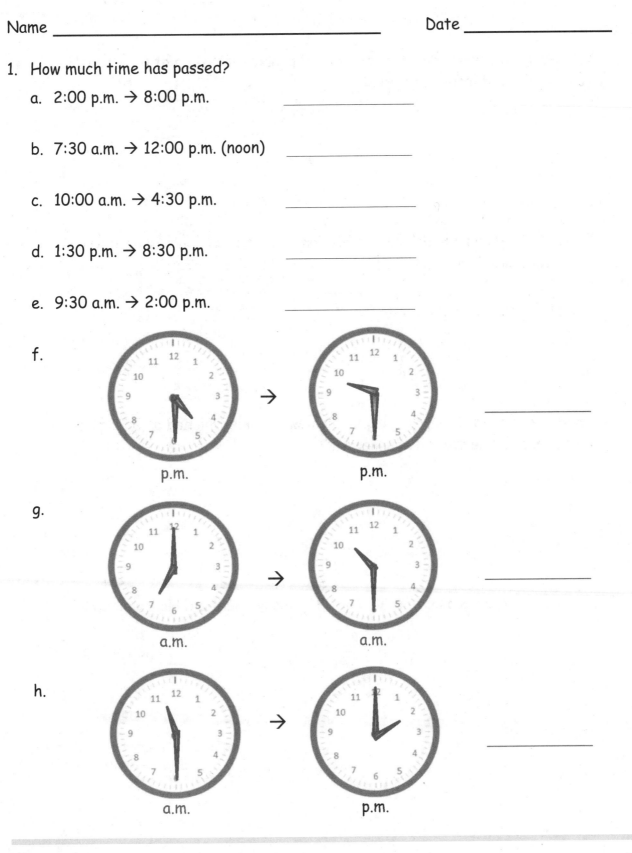

f.

p.m.   →   p.m.   _____

g.

a.m.   →   a.m.   _____

h.

a.m.   →   p.m.   _____

2. Solve.

   a. Kylie started basketball practice at 2:30 p.m. and finished at 6:00 p.m. How long was Kylie at basketball practice?

   b. Jamal spent 4 and a half hours at his family picnic. It started at 1:30 p.m. What time did Jamal leave?

   c. Christopher spent 2 hours doing his homework. He finished at 5:30 p.m. What time did he start his homework?

   d. Henry slept from 8 p.m. to 6:30 a.m. How many hours did Henry sleep?

Lesson 16:   Solve elapsed time problems involving whole hours and a half hour.

© 2018 Great Minds®. eureka-math.org

EUREKA MATH

Credits

Great Minds® has made every effort to obtain permission for the reprinting of all copyrighted material. If any owner of copyrighted material is not acknowledged herein, please contact Great Minds for proper acknowledgment in all future editions and reprints of this module.

- Module 7, Lesson 22, p. 180: Flathead screwdriver photo credit: Joao Virissimo / Shutterstock.com